PROTECTING NATURE, SAVING CREATION

ECOLOGICAL CONFLICTS, RELIGIOUS PASSIONS, AND POLITICAL QUANDARIES

Edited by

Pasquale Gagliardi

Anne Marie Reijnen

Philipp Valentini

palgrave
macmillan

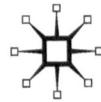

PROTECTING NATURE, SAVING CREATION
Copyright © Pasquale Gagliardi, Anne Marie Reijnen, and Philipp Valentini, 2013.

First published in 2013 by
PALGRAVE MACMILLAN®
in the United States—a division of St. Martin's Press LLC,
175 Fifth Avenue, New York, NY 10010.

Where this book is distributed in the UK, Europe and the rest of the world,
this is by Palgrave Macmillan, a division of Macmillan Publishers Limited,
registered in England, company number 785998, of Houndmills,
Basingstoke, Hampshire RG21 6XS.

Palgrave Macmillan is the global academic imprint of the above companies
and has companies and representatives throughout the world.

Palgrave® and Macmillan® are registered trademarks in the United States,
the United Kingdom, Europe and other countries.

ISBN: 978–1–137–36175–2

Library of Congress Cataloging-in-Publication Data is available from the
Library of Congress.

A catalogue record of the book is available from the British Library.

Design by Newgen Knowledge Works (P) Ltd., Chennai, India.

First edition: December 2013

10 9 8 7 6 5 4 3 2 1

CONTENTS

Part II   Afterthoughts

Figures and Table

Figures

Cover art *Noah sends out the dove from the ark*, mosaic, courtesy of "Procuratoria della Basilica di San Marco," Venice

Table

ACKNOWLEDGMENTS

Editing a work by various hands is not an easy task. It requires the creation of a temporary microorganization linking publisher, editors, and the various authors, so as to reconcile the various exigencies, styles, points of view with the need for something reasonably unitary in the outcome. The capable hand of Anna Lombardi held this undertaking on a steady course; her tenacity, her scrupulous efficiency, and her eye for detail have been of inestimable help to the editors in bringing this publishing venture into safe harbor.

PASQUALE GAGLIARDI
ANNE MARIE REIJNEN
PHILIPP VALENTINI

List of Contributors

Matthew Engelke is a Reader in Anthropology at the London School of Economics and Political Science. He has conducted extensive research in Zimbabwe and England on religion, and has particular interests in ritual, semiotics, secularism, and religious language. His book, *A Problem of Presence: Beyond Scripture in an African Church*, won the 2008 Clifford Geertz Prize and the 2009 Victor Turner Prize. His most recent research has been on humanists and atheists in Britain, looking in particular at nonreligious funerals. Matthew Engelke is the Honorary Editor of the *Journal of the Royal Anthropological Institute* and Editor of the Prickly Paradigm Press.

Pasquale Gagliardi, former Professor of Sociology of Organization at the Catholic University of Milan, is at present Secretary General of the Giorgio Cini Foundation in Venice. During the 1990s he contributed to the raising and development of the "aesthetics of organization" as a specific field of enquiry within organizational studies. His present research focuses on the relationship between culture, aesthetic knowledge and organizational order. He has widely published on these topics in Italian and English. Among his publications: *Le imprese come culture* (Torino: Isedi, 1986); *Symbols and Artifacts. Views of Corporate Landscape* (Berlin/New York: de Gruyter, 1990); *Studies of Organization in the European Tradition* (Greenwich, CT: Jai Press, 1995); *Narratives We Organize By*, coedited with Barbara Czarniawska (Amsterdam/Philadelphia: John Benjamins, 2003); *Management Education and Humanities*, coedited with Barbara Czarniawska (Cheltenham, UK: Edward Elgar, 2006); *Les atmosphères de la politique. Dialogue pour un monde commun*, coedited with Bruno Latour (Paris: Les Empêcheurs de penser en rond/Le Seuil, 2006); *Coping with the Past. Creative Perspectives on Conservation and Restoration*, coedited with Bruno Latour and Pedro Memelsdorff (Firenze: Leo S. Olschki, 2010); *Il gusto dell'organizzazione. Estetica, conoscenza, management* (Milano: Edizioni Angelo Guerini e Associati, 2011).

Eric Geoffroy is an expert in Islam and Professor in Islamic Studies in the Department of Arabic and Islamic studies at the University of Strasbourg (France). He also teaches at the Open University of Catalonia, and at the Catholic University of Louvain (Belgium). He is a specialist in the study of Sufism and sanctity in Islam. Among others, his research also extends to a comparison of mysticism, and to issues of spirituality in the contemporary world (spirituality and globalization; spirituality and ecology). He gives many lectures in the field of Sufism and more generally of Islamic culture all over the world (Europe, the Arab world, United States, Indonesia). So far, Eric Geoffroy has had seven books published. He is the author of numerous articles in magazines specialized in Islamology as well. Besides this, he has also participated in international conferences and made some contributions to key reference books (*Les voies d'Allah, Dictionnaire critique de l'ésotérisme, Dictionnaire du Coran, histoire de l'Islam et des musulmans en France du Moyen-Âge à nos jours*). Eric Geoffroy himself works at organizing conferences and seminars (Bibliotheca Alexandrina in Alexandria, European Council in Strasbourg, etc.). He has been selected for the Who's Who in the World 2011. Among his publications: *L'islam sera spirituel ou ne sera plus* (Paris: Le Seuil, 2009); *Le soufisme, voie intérieure de l'Islam* (Paris: Le Seuil, 2009). In English: *Introduction to Sufism* (Bloomington, IN: World Wisdom, 2010); *Une voie soufie dans le monde: la Shâdhiliyya* (Paris: Gnôsis-Éditions de France, 2011); *Jihâd et contemplation: Vie et enseignement d'un soufi au temps des croisades*(Paris:Albour aq,2000) .

Izabela Jurasz is specialized in Syrian Patristics, Greek Patristics, and ecumenical theology. She is a senior lecturer at the *Institut Catholique de Paris* (Theologicum—Faculty of Theology and Religious Sciences) and researcher at the *Centre Léon Robin* (UMR 8061, Paris IV). Among her publications: "La légende syriaque de l'invention de la Croix," *Doctrine d'Addai*: 16–30; "Implication politiques et théologiques," in *Les apocryphes chrétiens des premiers siècles. Mémoire et tradition*, edited by F.-M. Humann and J. N. Pérès (Paris: Éditions Desclée de Brouwer, 2009), 91–119; "Le 'Notre Père' commenté par Cyrille d'Alexandrie et ses disciples de la tradition non-chalcédonienne," in *Lire le Notre Père avec les Pères*, edited by D. Vigne (Paris: Parole et Silence, 2009), 319–341; "Unité de la Parole de Dieu dans le Diatessaron de Tatien," in *La Parole de Dieu dans le patrimoine syriaque. Au risque de la diversité religieuse et culturelle*. Patrimoine Syriaque. Actes du colloque XII (Antélias-Liban: Édition du CERO, 2010), 53–63; "Ecriture, création, révélation. Théologie de la Parole de Dieu dans les lectures patristiques du Prologue de Jean dans le milieu antiochien (IIe–IVe siècle)," in *La Parole de Dieu dans le patrimoine syriaque. Au risque de la diversité religieuse et culturelle*. Patrimoine Syriaque. Actes du colloque XII (Antélias-Liban: Édition du CERO, 2010), 199–214; "Mort et résurrection de l'âme chez Bardesane," *Chora. Revue d'études anciennes et médiévales*, 9–10 (2011–2012), 399–427.

Bruno Latour, born in 1947 in Beaune, Burgundy (France), from a wine grower family, was trained first as a philosopher and then an anthropologist. From 1982 to 2006, he has been professor at the *Centre de sociologie de l'Innovation at the Ecole nationale supérieure des mines* in Paris and, for various periods, visiting professor at the University of California, San Diego, at the London School of Economics and in the history of science department of Harvard University. He is now professor at Sciences Po in Paris where he is also the vice-president for research of that school. After field studies in

Africa and California he specialized in the analysis of scientists and engineers at work. In addition to work in philosophy, history, sociology, and anthropology of science, he has collaborated into many studies in science policy and research management. He has written *Laboratory Life* (Princeton, NJ: Princeton University Press, [1986] 1979); *Science in Action* (Cambridge, MA: Harvard University Press, 1987), and *The Pasteurization of France* (Cambridge, MA: Harvard University Press, 1988). He also published a field study on an automatic subway system *Aramis or the Love of Technology* (Cambridge, MA: Harvard University Press, 1996) and an essay on symmetric anthropology "We have never been modern." He has also gathered a series of essays, *Pandora's Hope: Essays in the Reality of Science Studies* (Cambridge, MA: Harvard University Press, 1999), to explore the consequences of the "science wars." After having directed several thesis on various environmental crisis, he published a book on the political philosophy of the environment *Politics of Nature* (Cambridge, MA: Harvard University Press, 2004). All of those have been translated in many languages. In a series of books, he has been exploring the consequences of science studies on different traditional topics of the social sciences: religion in *On the Modern Cult of the Factish Gods* (Durham, NC: Duke University Press, 2010) and *Jubiler ou les tourments de la parole religieuse* (Paris: Les Empêcheurs de penser en rond/Le Seuil, 2002) and social theory in *Paris ville invisible*, a photographic essay on the technical and social aspects of the city of Paris (available on the web in English *Paris Invisible City*). After a long field work on one of the French supreme courts, he has recently published a monograph *La fabrique du droit. Une ethnographie du Conseil d'Etat* (Paris: La Découverte, 2002) (also now in English). A new presentation of the social theory, which he has developed with his colleagues in Paris, is available at Oxford University Press, under the title *Reassembling the Social, an Introduction to Actor Network Theory* (2005). After having curated a major international exhibition in Karlsruhe at the ZKM (*Zentrum für Kunst und Medientechnologie*), *Iconoclash beyond the Image Wars in Science, Religion and Art*, he has curated another one also with Peter Weibel *Making Things Public. Atmospheres of Democracy* that has closed in October 2005 (both catalogs are with MIT Press). While in Sciences Po, he has created the *Médialab* to seize the chance offered to social theory by the spread of digital methods and has created, together with Valérie Pihet a new experimental program in art and politics (SPEAP). Having been awarded an ERC (European Research Council) grant to pursue an inquiry into modes of existence, he is now for three years engaged into the making of this collaborative digital platform.

Ignazio Musu is Professor of Economics and Environmental Economics at Ca' Foscari University of Venice. He is President of the Center on Thematic Environmental Networks of Venice International University. He is also member of Council of the Bank of Italy and various scientific and cultural foundations as the Foundation of Venice and European Association of Environmental and Resource Economists. He is corresponding fellow of the

Accademia Nazionale dei Lincei. He published many volumes on environmental economics and policy, and on sustainable development. Since 2003, he specialized in the problems of the sustainable development in China. He has been recently coeditor of the books *Sustainable Development and Environmental Management* (Berlin Heidelberg: Springer-Verlag, 2008) and *The Chinese Economy* (Berlin Heidelberg: Springer-Verlag, 2013).

Ted Nordhaus and **Michael Shellenberger** are leading global thinkers on energy, climate, security, human development, and politics. They are executive editors of the *Breakthrough Journal*. Their 2007 book *Break Through* was called "prescient" by *Time* and "the most important thing to happen to environmentalism since Silent Spring" by *Wired*. Their 2004 essay, "The Death of Environmentalism," was featured on the front page of the Sunday *New York Times*, sparked a national debate, and inspired a generation of young environmentalists. Over the years, the two have been profiled in the *New York Times*, *Wired*, the *National Review*, *The New Republic*, and on *NPR*. In 2007, they received the Green Book Award and *Time* magazine's 2008 "Heroes of the Environment" award. In 2011, Nordhaus and Shellenberger started the *Breakthrough Journal*, which *The New Republic* called "among the most complete answers" to the question of how to modernize liberal thought, and the *National Review* called "The most promising effort at self-criticism by our liberal cousins in a long time." Shellenberger and Nordhaus are leaders of a paradigm shift in climate and energy policy. They proposed "making clean energy cheap" in the *Harvard Law and Policy Review*, explained why the Kyoto climate treaty failed in *Democracy Journal*, and predicted the bursting of the green bubble in the *New Republic* and *Los Angeles Times*. The two predicted the failure of cap and trade in the *American Prospect*, criticized "green jobs" in the *New Republic*, and pointed a way forward for climate policy in the *Wall Street Journal*. The two have written on intellectual property for *Slate*, counterterrorism for *Roll Call*, the end of the war on terror for the *Atlantic*, and modernization as a new ecological theology for *Orion*. Ted Nordhaus is a graduate of the University of California, Berkeley, and Michael Shellenberger is a graduate of Earlham College and holds a masters degree in cultural anthropology from the University of California, Santa Cruz. The two live in the Bay Area and travel widely.

Anne Marie Reijnen is an ordained minister and a Protestant theologian. She currently holds the Kairos Chair of the ISEO (Institut Supérieur d'Etudes Oecuméniques/Paris Institute for Advanced Studies in Ecumenism) of the Catholic University Paris (Institut Catholique de Paris). She is also a University Professor of the FPG/FUTP (Faculté Universitaire de Théologie Protestante) in Brussesl, Professor of Theology in Brussels and Paris. She is a member of the Center for Theological Inquiry (Princeton, NJ), where she was in residence in 2002, 2005, and 2008. Since 2010 she participates in the ecumenical *Groupe des Dombes*. Her fields include: relations between Christianity and Judaism; Christology; ecology and theology (e.g., the dignity of non-human animals); the work of Paul Tillich; feminist theology; African American theology; and spirituality. A selection from her publications: *L'Ombre de Dieu sur terre. Un essai sur l'incarnation* (Genève: Labor et Fides, 1998); *L'Ange obstiné. Ténacité de l'imaginaire spiritual* (Genève: Labor et Fides, 2000); "Das Heilige als Kategorie bei Rudolf Otto und Paul Tillich," in *Mystisches Erbe in Paul Tillich*, edited by G. Hummel and D. Lax (New York: de Gruyter, 2000); "Maître ou parasite? Habiter la terre en toute conscience," *Revue Théologique de Louvain* no. 1 (2000): 169–189; "Holy impatience. Participating in the redemption of the world," in *Théologie et culture. Hommage à Jean Richard*, edited by M. Dumas, F. Nault, and L. Pelletier (Laval: Presses de l'Université Laval, 2004); "Tillich's Christology," in *The Cambridge Companion to Tillich*, edited by R. Re Manning (Cambridge: Cambridge University Press, 2008); "Contre Leibowitz: Les origines juives du christianisme," *Cités* 34 no. 4 (2008); "Le 'cas' de l'animal," in *La dignité aujourd'hui*, edited by A.-M. Dillens (Bruxelles: Publications des Facultés Universitaires Saint-Louis, 2008); "Sexes, genres et genre humain: un itinéraire théologique," in *Le Christianisme est-il misogyne? Place et rôle de la femme dans les Eglises*, edited by J. Famerée (Bruxelles: Lumen Vitae, 2010); "Protestantism as 'Gestalt' in Tillich's analysis of culture," in *Paul Tillichs Theologie der Kultur*, edited by C. Danz and W. Schuessler (Berlin/Boston: Walter de Gruyter, 2011), 279–292; "Protestants and Jews after the Shoah: confessing God and the Son of God amidst the shambles of European history," in *Thinking the Divine in Interreligious Encounter*, edited by N. Hintersteiner and F. Bousquet (Amsterdam/New York: Rodopi, 2012), 127–146; the articles "Hommes/Femmes" and "Judaisme," in *Dictionnaire d'Ethique Chrétienne* edited by Laurent Lemoine and Denis Mueller (Paris: Cerf, 2013), 1082–1090 and 1162–1170.

Simon Schaffer is Professor of History of Science in the University of Cambridge. He studied at Cambridge and Harvard and taught at Imperial College London. With Steven Shapin, he coauthored *Leviathan and the Air Pump: Hobbes, Boyle and the Experimental Life* (Princeton, NJ: Princeton University Press, 1985). He coedited *The Sciences in Enlightened Europe* (Chicago: University of Chicago Press, 1999), *The Mindful Hand* (Amsterdam: KNAW, 2007), and *The Brokered World* (Sagamore Beach, MA: Science History Publications, 2009). He helps direct a research project on the history of the Board of Longitude, 1714–1828, and sits on the advisory boards of the Science Museum, London and of the Arts and Humanities Centre at the Natural History Museum, London.

Angelo Scola, having obtained doctorates in philosophy and in theology, from 1982 to 1995, was Professor of Theological Anthropology at the Pontifical John Paul II Institute of the Pontifical Lateran University. In 1991 he was appointed Bishop of Grosseto. From 1995 to 2002 he was Rector of the Pontifical Lateran University and President of the Pontifical John Paul II Institute. He was made the Patriarch of Venice on January 5, 2002 and a Cardinal on October 21, 2003. Since June 28, 2011 he has been the Archbishop of Milan. His most recent books include *Uomo-donna: il "caso serio" dell'amore* (Milano: Marietti, 2002), Capri Prize 2003; *Eucaristia: incontro di libertà* (Siena: Cantagalli, 2005); *Chi è la Chiesa? Una chiave antropologica e sacramentale per l'ecclesiologia* (Brescia: Queriniana, 2005); *Una nuova laicità. Temi per una società plurale* (Venezia: Marsilio, 2007); *Come nasce e come vive la comunità cristiana* (Venezia: Marcianum Press, 2007); *Dio? Ateismo della ragione e ragioni della fede* (Venezia: Marsilio, 2008); *Maria, la donna* (Siena: Cantagalli, 2009); *Buone ragioni per la vita in comune* (Milano: Mondadori, 2010); *Vivere da grandi* (Venezia: Marcianum Press, 2011); *La vita buona* (Milano: Mondadori, 2012).

Elizabeth Theokritoff is an Orthodox Christian independent scholar and theological translator from Greek, with particular interests in theology of creation and in liturgical theology. She has lectured widely, including serving as a visiting lecturer at the Institute for Orthodox Christian Studies, Cambridge, and has been involved in various conferences and scholarly projects connected with theology of creation. Publications include "Creation and priesthood in modern orthodox thinking," *Ecotheology* 10.3 (December 2005): 344–363; "Creator and Creation," in *Cambridge Companion to Orthodox Christian Theology*, edited by Mary Cunningham and Elizabeth Theokritoff (Cambridge: Cambridge University Press, 2008); "Crise écologique et témoignage chrétien: défi pour l'Église," *Contacts* 227 (July–September 2009): 251–268; *Living in God's Creation: Orthodox Perspectives on Ecology* (Yonkers, NY: St. Vladimir's Seminary Press, 2009); "Cosmic priesthood and the human animal: speaking of man and the natural world in a scientific age," in *Thinking Modernity: Towards a Reconfiguration of the Relationship between Orthodox Theology and Modern Culture*, edited by Assaad Elias Katan and Fadi A. Georgi (Balamand: St. John of Damascus Institute of Theology, 2010); "The High Word's mystery play: creation and salvation in St Maximus the confessor," in *Creation and Salvation*, vol. 1: *A Mosaic of Selected Classical Christian Theologies*, edited by Ernst M. Conradie (Münster: LIT Verlag, 2011).

George Theokritoff has degrees in Geology and Paleontology from the University of London. His master's thesis was on parts of Galway and Mayo, Ireland, and his doctoral dissertation was on parts of eastern New York State and west-central Vermont. He has taught at colleges and universities in England, Canada and the United States and retired as Professor Emeritus from Rutgers University in New Jersey where most of his teaching was concerned with Earth History, the History of Life, and Paleontology. He is also author or co-author of many research articles in peer-reviewed journals. More recently, he has become interested in the interface of science to traditional Christianity.

Philipp Valentini studied religious sciences at the University of Venice where he wrote his master degree research under the direction of Angelo Scarabel, on the theme *Nuove prospettive sulla storia del sufismo in Africa occidentale* (2004–2007). He then worked three years at the *Fondazione Giorgio Cini*, Venice, where he contributed to the planning and organization of three *Dialoghi di San Giorgio: Landscaping Politics* (2009), *Protecting Nature or Saving Creation? Ecological Conflicts and Religious Passions* (2010), *Myths of Universal Knowledge and Aesthetics of Global Imaging* (2012), and to the editing processes of the books drawn from them. In Venice, he also helped in conceiving and launching a new center dedicated to the comparative study of spiritualities and cultures (http://www.cini.it/en/fondazione/institutes-and-centres/civilta-e-spiritualita-comparate) (2012). He actually works at the University of Fribourg as an assistant to Helmut Zander who teaches Comparative Study of the History of Religions and Interreligious Dialogue (http://www.unifr.ch/screl/de/team). His PhD dissertation deals with the ways in which the European culture (René Guénon and Henry Corbin) have worked out an understanding of Sufism (Ibn ʻArabi in particular) heavily influenced by the hermeneutical frame conceived by Franz Joseph Molitor (1779–1860) for understanding the Jewish tradition.

Andrea Vicini, SJ, is Associate Professor of Moral Theology at Boston College School of Theology and Ministry (Massachusetts, USA). Medical doctor and pediatrician, his theological training includes a Bachelor in Theology (*Centre Sèvres, Paris*), a Licentiate in Sacred Theology (Weston Jesuit School of Theology), a PhD in Theological Ethics (Boston College), and a Doctorate in Sacred Theology (Faculty of Theology of Southern Italy). He has taught in Italy, Albania, Mexico, Chad, and France. Lecturer and member of important associations of moral theologians and bioethicists (in Italy, Europe, and in the United States), his research interests include fundamental moral theology, bioethics, biotechnologies, medical ethics, sexuality, and environmental issues. Among his recent publications are *Genetica umana e bene comune* (*Human Genetics and the Common Good*) (Cinisello Balsamo: San Paolo, 2008) and its translation: *Genética humana e bem comum* (São Paolo: Loyola-São Camilo, 2011); "Bioethics: basic questions and extraordinary developments," *Theological Studies* (2012); "Ethical challenges of genetics today: from the lab, through the clinic, to the pews," *Studia Moralia* (2012); "Imaging in severe disorders of consciousness: rethinking consciousness, identity and care in a relational key," *Journal of the Society of Christian Ethics* (2012); "La loi morale naturelle: Perspectives internationales pour la réflexion bioéthique contemporaine," *Revue de l'Institut Catholique de Paris-Transversalités* (2012); "New work in ethics of environment and sustainability," *Asian Horizons: Dharmaram Journal of Theology* (2012).

Eduardo Viveiros de Castro is an anthropologist who has conducted research in Brazilian Amazonia, most of it among the Araweté of the Middle Xingu. He was Simon Bolívar Professor of Latin American Studies in the University of Cambridge (1997–1998) and *Directeur de Recherche* at the CNRS (Centre National de la Recherche Scientifique) in Paris (1999–2001).

His publications include *From the Enemy's Point of View: Humanity and Divinity in an Amazonian Society* (Chicago: University of Chicago Press, 1992); *A Inconstância da Alma Selvagem e Outros Ensaios de Antropologia* (São Paulo: Cosac Naify, 2002), and *Métaphysiques cannibals* (Paris: Presses Universitairesd e France,20 09).

INTRODUCTION

Pasquale Gagliardi

THE BEGINNING OF AN ADVENTURE

This book tells the story of an intellectual adventure that involved scholars and experts from various disciplines and of different religious beliefs. The adventure reached a climax in a three-day meeting on the island of San Giorgio Maggiore, Venice, in September 2010 but has actually stretched out over a period of three years, from summer 2009 to summer 2012.

Everything began on a glorious afternoon in late July 2009 with a conversation between Bruno Latour and myself. For some years Latour had been collaborating on a new cultural initiative set up by the Giorgio Cini Foundation in 2004: the *Dialoghi di San Giorgio*, a series of multidisciplinary meetings aimed at encouraging dialogue and debate on key issues for contemporary society, held every year on San Giorgio in mid-September.[1] Latour and I were looking for a theme for the 2010 Dialogue. We had just revisited a miracle of faith, art and technique: the stories from the life of St. Francis in the chapels of the Sacro Monte di Orta. Now, before our eyes a stunningly beautiful landscape unfolded: opposite were the snowfields of the Monte Rosa, and below the still blue mirror of Lake Orta framing the Romanesque architecture on the island of San Giulio as if it were a gray pearl. Our conversation touched on the beauty and fragility of the Earth, on ecological conflicts and the difficulty of implementing effective environmental policies. Then, suddenly, paraphrasing a verse from the Gospel according to St. Mark,[2] Latour said: "What shall it profit a man to save his soul, if he loses the Earth?"[3] This is what gave rise to the idea of devoting a *Dialogo* to the relationship between ecology and theology.

The first step was to select an initial group of scholars to be invited to take part. They were then sent an informal note concisely illustrating the idea to sound out its appeal. We received enthusiastic responses and many comments. The comments of each person were in turn sent to all the other potential participants. New names were put forward and a remote dialogue, so to speak, gradually got underway. This initial group work was useful in constructing the intellectual framework for the *Dialogo*, which was later used to explain the subject and the reasons for the meeting both to newly invited participants and the general public.

THE TOPIC AND REASONS FOR THE DIALOGO

The urgency of the *Dialogo* lay in the widely shared awareness that the gamut of passions mobilized by ecology so far has not reached the level or intensity required for the huge task facing humanity today concerning the fate of the Earth. In the past religion seems to have been able to mobilize transformative passions and energies that produced radical change—of a social, cultural, and even physical nature—on an extraordinary scale. This was arguably because religious passions—for better or for worse—encourage transcendental experience through which human beings become detached from earthly things and can act on them with greater freedom.

The basic question we started from was as follows: can religion help us tackle the ecological crisis we are now facing? The urgency to give spiritual and religious depth to the ecological question appears more evident, if we define the ecosystem not only as a physical environment to be preserved but also as an intrinsically cultural space: the deforestation of the Amazon is not only synonymous with the massive destruction of the forests and the extinction of animal species but also with the destruction of social and mental environments.

There is no easy answer to the basic question. It is not clear if an equivalent level of energy is still available and, if it is, whether it can be used for ecological ends. Attempting to give an answer to the question means exploring the relation between ecology and theology. The debate on the relation between theology and ecology is nothing new. The striking aspect is that the already immense literature on the subject seems not to have an explicitly significant impact on ecological policies. To our mind, there are at least three explanations for the relative sterility of the debate.

First, the fact that in general the various communities—theologians, ecologists, economists, sociologists, and business people—discuss the question internally without any real interaction between the various communities (and this explains the difficulty in formulating a political agenda).

Second, far too often the debate is based on an outmoded conception of science, on a lack of discrimination between the notions of nature, the creation, and the cosmos and on debatable notions of religion, especially Christianity. Reopening the debate on the relation between science and religion—a notoriously trite topic—implies exploring the tension between Nature and the Creation by referring to the ancient theologies elaborated by the Early Church Fathers but also the various natural theology traditions.

Third, what is usually completely overlooked in any analysis of the relation between ecology and theology is the role of conflicts and passion. Many authors seem to presuppose that the two fields are naturally and harmoniously linked, when unfortunately both Nature and the Creation have no lack of drastically conflicting dimensions. As an icon for our dialogue we chose a mosaic in St. Mark's showing the idyllic scene of Noah releasing a dove, the symbol of the pacified soul; but in the foreground on the endless blue sea, a raven—an allegory of the restless bodily soul—is devouring a carcass. In the Epistle to the Ephesians, St. Paul writes: "Above all, taking the shield of faith, wherewith ye shall be able to quench all the fiery darts of the wicked. And take the helmet of salvation, and the sword of the Spirit, which is the word of God." With this warlike image he highlights the conflictual dimension of salvation.

Another limit in the contemporary eco-theological debate is that it concentrates on "saving" the planet but overlooks the semantic pregnancy of the notion ("salvation") imbued with religious connotations that go much further than a simple question of survival (this oversight arguably reveals the anguish informing the desire of those not wishing to save anyone other than themselves). What then does "salvation" mean? Who and what must we save and why? Is it possible to give an answer to this question without radically redefining the concepts of nature, environment, ecosystem, cosmos and their relation with soul, society, and the divine?

The key question is therefore perhaps how to mobilize the notions, cosmologies, and rituals characterizing some religious traditions, provided, however, they do not overlook the conflicts underlying the ecological debate and the essential role of politics: without an adequate consideration of the conflicts, arguably no ecological policy is possible.

THE METHOD

The participants, who had various roles at the *Dialogo*, are all those who appear as contributors to this book: theologians of various confessions, environmental strategists, economists, philosophers, historians, sociologists, and anthropologists. The invited experts were not asked to give lectures in the traditional sense of the term but to take part in conversations with all the attendant risks. We were convinced that an exchange of different disciplinary and cultural points of view has the potential to generate more fertile ideas and intuitions than other more traditional academic formats of debating and communicating knowledge. This comparative exercise appeared to the *Dialogo* organizers as the most significant and perhaps the only antidote possible to all kinds of rampant fundamentalisms. It was also a resolute way of being faithful to the unarmed St. George on the dome of the church on the island of San Giorgio, to which history and perhaps also nature have assigned the role of being a place for meetings and exchanges between disciplines and cultures.

Each participant was asked to send or indicate a previously published paper that addressed significant aspects of the chosen topic. These writings were then sent to all the other participants, not necessarily as an anticipation of what each would have said, but as a way of introducing oneself to the others and to improving mutual knowledge ahead of the three days of meetings so as to make them more fruitful.

THE EXPERIENCE

The evening before the first day of the *Dialogo* (September 13, 2010), an inaugural event was held. It consisted of two "speeches" framed by the performance of two pieces of music. In keeping with the Cini Foundation tradition, we wished that the eminently intellectual experience of the *Dialogo* was introduced and supplemented by an aesthetic experience, stimulating forms of cognition not involving logical-analytical processes but intuition, emotions, and feelings.

The first piece of music was Milhaud's *The Creation of the World*, a composition inspired by the African myths of the Creation evoking the chaos before the world, the appearance of animals and vegetation, and the birth of man and woman. Pasquale Gagliardi, Secretary General of the Cini Foundation, then presented the program of the *Dialogo*. Next Cardinal Angelo Scola, Patriarch of Venice, pronounced an allocution on the theme *Taking in the Real: Human Beings and the Earth* (see Chapter 1 in this book). The event ended with the three movements from Mahler's *Song of the Earth* performed by the alto soloist, including the outstanding—not only because of its length but also for its symbolic value—last movement of the whole composition ("Farewell").

The *Dialogo* was held over the next three days, from September 14 to 16. The participants sat for three days at the round table in the *Sala dei Cipressi* in the Cini Foundation. Each day, three debate sessions were held from 9 a.m. to 4 p.m. The proceedings were mainly conducted in English and each session was introduced by one of the participants. The general public could attend the sessions without taking part in the discussion directly. The organizers wished that the public should only be a silent presence to underscore and enhance that "listening atmosphere," deemed to be of fundamental importance for the success of the event.

The sequence of the chapters in this book reflects the order of the sessions introduced by the various participants and fairly faithfully documents what happened during the three days in the recorded formal sessions open to the public. On each of the three days, however, the participants also met informally at the end of the sessions—from 4 to 5 p.m.—in the living room of the Foundation guest quarters for a kind of "writing workshop." Their aim was to set down some of the main ideas that had been thrown up during the day in order to draft gradually over the three days a kind of charter or manifesto, which could then be handed out at the end of the meeting.

Unsurprisingly, different and contrasting opinions immediately emerged about the aims and methodology of the workshop. Marked tensions between the views of different participants were expressed, and are clear in the record of the debates. There was no provision for a final session in which to draw conclusions or exchange closing remarks. Rather, a draft manifesto was fiercely debated and, despite attempts at mediation, participants were unable to reach an agreed text. Such dissent might perhaps have been expected: conflict and passion provided not only the salutary theme but also the structural model for our dialogue. Having discussed the matter at length with my coeditors, I was reminded of the clue suggested in 1985 by Thompson and James[4] in their analysis of debates on environmental policies, taken up again in 1992 by Mary Douglas. Applying a cultural theory to these often heated, confused, and inconclusive debates reveals that the different positions and arguments are based on various (robust, fragile, or unpredictable) myths of nature:

> Cultural theory starts by identifying the context of appeals to nature, then it uncovers the strategies of debate, and shows the foundation myth as the final clinching argument. In fact, the base line does not clinch anything, because there is no way of demonstrating that one or the other myth of nature is the right one. At some point the summoning of evidence becomes unnecessary, more evidence will not settle the divergence of opinion. Somewhere along the line the debaters realize that they are facing infinite regress, more explanations calling forth more counter-explanations, and when this happens, theorizing has to end. In a debate about what to do with the environment, explanations come to rest on their appropriate myths of nature.[5]

We were left with a paradox. Despite, or indeed because of, the amount of energy devoted to reaching a consensual outcome, no formal conclusion was achieved: this was unprecedented in the history of the *Dialoghi*. This feeling of incompleteness, expressed by various participants, led me a few months after the event to ask everyone to make a retrospective interpretation and sum up their personal thoughts on the whole experience.[6] Most of the participants agreed to take part in this kind of virtual final session and their reflections are contained in the "Afterthoughts," the second part of the book.

NOTES

1. The *Dialoghi di San Giorgio* are a new version of a long-standing annual event at the Foundation, the *Corso di Alta Cultura* ("Course of High Culture"). For almost 50 years this course saw authoritative scholars and leading witnesses of our age come to the island of San Giorgio Maggiore, where the Foundation is based. The *Dialoghi* have gathered and developed the legacy of those courses by adapting them to the times without sacrificing their spirit and function. As at the time of the *Corsi di Alta Cultura*, today for the Cini Foundation the *Dialoghi* provide an emblematic opportunity to bear witness to the values that have always inspired its actions over time: a faith in knowledge that stems from dialogue, the search for truth in freedom, and the sense of providing a service to the scientific community, to the city and society in general. The *Dialoghi* of 2004 (*Atmospheres of Freedom. For an Ecology of Good Government*), 2005 (*The Architecture of Babel. Creations, Extinctions and Intercessions in the Languages of the Global World*), and 2007 (*Inheriting the Past. Tradition, Translation, Betrayal, Innovation*) led to the publication of the following books, respectively: Bruno Latour and Pasquale Gagliardi (eds.), *Les atmosphères de la politique. Dialogue pour un monde commun* (Paris: Les Empêcheurs de penser en rond/Le Seuil, 2006); Paolo Fabbri

and Tiziana Migliore (eds.), *The Architectures of Babel. Creation, Extinctions and Intercessions in the Language of the Global World* (Firenze: Leo S. Olschki, 2011); Pasquale Gagliardi, Bruno Latour and Pedro Memelsdorff (eds.), *Coping with the Past. Creative Perspectives on Conservation and Restoration* (Firenze: Leo S. Olschki, 2010).

2. "For what shall it profit a man, if he shall gain the whole world, and lose his own soul?" (Mark 8:36).

3. This phrase was actually the title of an article Bruno Latour had recently published: B. Latour, "Si tu viens à perdre la Terre, à quoi te sert d'avoir sauvé ton âme?," *Revue-Theologicum.fr*, http://www.catho-theo.net/spip.php?article248

4. M. Thompson and P. James, "The cultural theory of risk," in *Environmental Threats: Reception, Analysis, and Management*, edited by J. Brown (London: Belhaven, 1985).

5. M. Douglas, "In defence of shopping," in *Produktkulturen. Dynamik und Bedeutungswandel des Konsums*, edited by R. Eisendle and E. Miklautz (Frankfurt/New York: Campus Verlag, 1992).

6. See "A Preliminary Notice," p. 223.

Ariadne and the Minotaur: A Thread Winding through the Labyrinth

A Guide for Readers

Anne Marie Reijnen

Readers of this book will find several visions and perspectives on the topic of ecology, on religion and faith, and the political quandaries involved in "saving" the earth and the well-being of its inhabitants. Take a kaleidoscope: a finite number of fragments combines into a great variety of successive "images," depending each time on who happens to be "slanting" the instrument. Likewise, images are conjured up by the participants of the Dialogue in the pages you are about to read: ill-fated Pandora and her box; the engineer Victor Frankenstein and the monster who escapes from his laboratory (Figure 5.1); the Amerindian's Amazonian forest full of "human" jaguars; the "cosmic" vision of the planet as a "blue orange" (Figure 7.1); the city of Venice itself, saved from the flood by human ingenuity; Noah in the Ark, cupping his hand for God's dove; the bread and the cup of wine of the Lord's supper; the watchmaker who eyes his creation; Shakespeare's Caliban, and Macbeth. The Dialogue is not unlike Grayson Perry's contemporary tapestry *Map of Truths and Beliefs* (Figure 2.1). For all their passionately held differences, the participants share at least the conviction that their desire to confront ecology and theology will have political consequences. The question recurred: "who can save?"

As Franz Rosenzweig wrote, "Saving is the coming together of faith and life."

We now unroll a thread to navigate Ariadne's labyrinth, to avoid being waylaid by the Minotaur and its dappled offspring.

The book is divided into two parts: to start, the proceedings of the first "act" of the 2010 Dialogue: the written versions of the short papers delivered by the participants, followed by the transcription of the ensuing discussions. The opening speech by Cardinal Angelo Scola, the Patriarch of Venice, and the recital of Mahler's *Lied von der Erde* raised the curtain. The Patriarch offered a thoughtful preliminary: "Nature [...] is not only a set of 'things' but also of 'meanings' through which human freedom is called on to realize its own original vocation in the search for the face of the Creator" (p. 000). This first act was already preceded by a virtual dialogue prior to the actual meeting; during approximately a year, papers were exchanged among the participants.

The second part shows a majority of the participants continuing the process well after leaving the island of San Giorgio, thus enacting a second "act" of Dialogue. Some of these self-reflective fragments offer glimpses of shifts in perception by the authors. The Dialogue was a time both for friendship and for staunch conflict, at times for scuffles from entrenched positions. Was it a failed experiment? We believe the experiment was successful. Measuring the distances between our respective worldviews was the epistemological premise, the indispensable first step, into the labyrinth, where words are not what they seem *prima vista*: for example human, creation, saving, God(s), "nature" ... An introductory "patchwork" of statements by the participants, gathered by the editors, provides a sample of the Dialogue. Let me now provide a brief "guide for readers" of this book.

PART I: PROCEEDINGS OF THE DIALOGUE

The first papers address the *historical* dimension, by describing epochal moments of breakthrough, from early Christianity (Jurasz) to the British philosophical debate about "nature" in the eighteenth century (Schaffer) to the contemporary ecological crisis (Latour). Then comes a *confessional* sequence that is diachronic, critical, and self-critical: the participants analyze the visions, traditions, and rituals of the Orthodox faith (E. Theokritoff), the theology of Protestant churches (Reijnen), the teachings of the Catholic Church (Vicini), and finally some specific Muslim views, found in Sufism and in classical Coranic Islam (Geoffroy).

The third sequence takes into account *space*: the conflictual dynamics of different worldviews that coexist in a give place: between Africa and the "West" on African soil (Engelke); Amazonian "perspectivism" as the radical undermining of the traditional pattern of "object versus subject" (Viveiros de Castro).

The fourth and final sequence, outlining the contours of a *political ecology* for today, mobilizes empirical data: an economist's view (Musu) and an environmental strategy that relies on development and research of new technologies (Nordhaus and Shellenberger).

To summarize the findings of the eleven papers:

1. Izabela Jurasz, a Polish scholar of Syriac and Patristics from Paris, studies the alternative of "cosmos versus creation" in her contribution on the Syriac Fathers of the Church; she immediately problematizes the putative opposition, which is absent from Christian Antiquity. An adequate

methodology is required to approach the ancient texts: the proper hermeneutic stance is not to assume immediate understandings of the concepts used in Patristics, such as *ktisis, poiesis,* and *phusis.* About the first of the three terms (ktisis), Jurasz points out that it refers more to the Creator than to creation or creatures (or to "creativity"). The early Christians were wary of *poietes,* because of the suspicion that speaking of artifacts gets one too close to "make-believe." The interpretation of the third term, "nature" was as complex as it is today; indeed, it was never an unequivocal category. Somewhat better informed about the nuances of usage of the three terms, the reader is empowered to think about implications for the contemporary debate. God is not only called Creator; often, Church Fathers tie this in with *pronoia,* understood as benevolent and critical providence. Furthermore, the belief in the Creator implies the trust in the creative power of God's Word, and in divine creative Wisdom. The latter is a feminine agency: "Wisdom builds the universe so as to live with her creatures." To summarize: in Patristics, nature and creation are not simply interchangeable. The remit therefore is not to "protect nature in its present state," but to learn to see it as creation.

2. "Dame Nature cares nothing for us" sets the tone of Simon Schaffer's purview. The author is a British historian of sciences; he borrows the provocative statement from the radical journalist Richard Carlile; the implication being, again in the words of Carlile, that "we must take care of ourselves in spite of her." That is the function of the market; indeed it, along with production, survival, and mobilization are key notions replaced within their genealogy by Schaffer. Can one attribute them to an "invisible hand," seen with awe or optimism? Schaffer revisits the space "between" the problematic entities of "nature" and "society," a space that is socialized by moral economy. Where nature is stoic, regulation and proactive human agency is called for. But the stepmotherly indifference of nature can also become an instrument for colonial powers to deny the negative effects of their (agricultural) policies. Hence the critique: "why blame poor Nature when the fault lies at your own door?" The land as "creator" of human survival, at least of a limited population, is a problem discussed by Schaffer through Defoe's hero Robinson Crusoe and of the historical figures of Malthus and Paley. Schaffer's conclusion is that "Nature's seeming stoicism in the face of human suffering has long motivated admirable forces for survival and liberation."

3. Bruno Latour, French sociologist, in his contribution called "Creation and Salvation" sets out to map controversies. There is controversy or conflict regarding the Christian belief in continued creation, and the current rhetoric of "limits" and "downscaling" (*décroissance*). Mary Shelley's Gothic novel Frankenstein serves to illustrate Latour's point: the engineer Victor Frankenstein is to blame not for creating the monster but for abandoning it when it has escaped from the scientist's laboratory. This shows a lack of "care"; care as the willingness to attend to all and any "unintended consequences" that follow from the inventions of *homo faber.* Contrary

to a popular misconception, science should invest in the "faraway," while the Christian *logos* has us focus on the down-to-earth, on *le prochain*. On a far less consensual note, Latour pleads for the "intrusion" of Gaia, the personification of the Earth first theorized by James Lovelock. Finally, among the words at stake in the "diplomacy," the negotiations necessary to the dialogue between ecology, the sciences, and Christian theology, "apocalypse" in its different meanings must be watched closely.

4. Elisabeth Theokritoff, an independent Orthodox scholar, invites her audience and readers to consider both the "harmony" of the cosmos and the "imperfection" of human beings, notions familiar to Orthodox Christians but to a large extent "countercultural" in current society. The harmony of all that is visible and invisible can only be understood christologically: "the entire universe is destined to be renewed and transformed into Christ," just as human beings will share in the resurrection. Against both the pessimistic view of human beings prevalent in Western Christianity and against the "secular" unbridled belief in the human prowess, the Orthodox view holds that human beings have a glorious calling, and that they are "fallen." Their grand destiny is achieved by serving, not dominating; and specifically, through praise in worship. This is the true "nature" of humanity; nature thus implies the "intended state of persons and of objects." Nonhuman animals and other organisms "serve God's will," in accordance with their ordained nature; only human beings fail to do so. Saving creation cannot be achieved by a "conversion" to green activism.

5. "To contemplate the cosmos and to transform the world" is the dual task required of Protestantism, according to Anne Marie Reijnen, a Protestant theologian. The believers who interpret Christianity through the lens of the Reformation of the sixteenth century have gingerly espoused the task of transforming the world. Man is called to work as the bird to fly, according to Luther. The reticence to find holiness in nature (the need to combat paganism and idolatry) and the almost exclusive attention to redemption as an event in history have contributed to a certain "acosmism." Also, hearing is valued more highly than seeing. The starting point for Reijnen's paper goes against the grain of the tradition she represents: it is the vision of the "blue orange," the picture of the Earth beamed back to through space. This has become a visual commonplace and can become a common ground for Christians and non-Christians to work to preserve the fragile balance of our common habitat.

6. "Care" is the key notion for Andrea Vicini (SJ). He demonstrates the successful inculturation of the Aristotelian ethics of virtue in the broad tradition of the Catholic Church. Human beings as the "virtuous collaborators" of God learn to exercise "care" toward an ever expanding reality. The common good encompasses not only the current generations but also the generation yet unborn; human beings, but also all other creatures. Stewardship of the earth and kinship with all living creatures are put forward in the official documents of the Catholic Church, from 1975

onward. Vicini starts his contribution by highlighting the plight and the fight of the I-Kiribati, the people who inhabit the Kiribati islands in the Pacific Ocean: they are losing ground literally because of rising sea levels, a direct consequence of global warming. Ethical (virtuous) responses must influence particular situations, and concretely the people involved, and have global implications, that is with the "common good" in mind.

7. Eric Geoffroy, scholar of Islam, analyzes the "cosmism" of the Muslim tradition. All of creation is endowed with life because the creation stems from the "Ever Living" (one of the many names of Allah). The unity of the cosmos reflects the divine Oneness (*Tawhîd*). The Divine can be said to be immanent in matter because the divine Mercy (*Rahma*) "encompasses all things" (*Sura* 7, 156). Things are of course differentiated; *Sura* 114 inventorizes the different "levels," starting with the astral realm to the human realm, and in between the mineral, vegetal, non human animal, the realm of invisible beings (*djinns*). According to the great Sufi poet Rumi (thirteenth century CE), human beings carry in themselves the life of all these realms. Human beings are the representatives of the Divine on earth, having received freedom and responsibility, yet beset by corruption and pride. Geoffroy underlines an alternative understanding of *Sharia*, commonly misrepresented in the Western media as the rationale for aggressive proselytism; it is better understood as a Cosmic law, related to the Sanskrit word *Dharma*; its five tenets are respect for human life, religion, reason, procreation, and private property. No single Muslim state has thus far achieved all of these goals; an Islamic "liberation theology" is called for, to enable the spiritual disentanglement from consumerism.

8. Matthew Engelke "internalizes" conflict in the paper given from his ethnographical perspective. There were underlying and explicit conflicts in the previous papers and debates, but in Engelke's account, conflict becomes one of the topics; conflicting desires and commitments are played out in ecology and economy, in the encounter between the West and the "global South." The colonial enterprise also left a mark on the landscape, for example, the trees from the Scottish highlands that took root in "Rhodesia" (Zimbabwe). Whether that particular instance of artificial "greening" was positive or negative must remain undecided. The main thrust of Engelke's contribution is the question whether ecology is giving a language of universalism to the "traditional religions" that allows them to become consonant with Christianity. There were in fact tribal "eco-theologies" well before the current (partial) "greening" of West, in the political realm and in the theology of the three monotheist religions.

9. The contribution of Eduardo Viveiros de Castro, anthropologist, makes two fundamental points. The first has to do with worldviews. The Amazonian Indian has a radically different perspective on the human realm and the other realms; this difference makes it impossible to situate them within what we call "modernity": premodern, modern, late modern...In

a sense, according to Viveiros de Castro, Amazonian Indians are neither in nature (contrarily to what is widely believed), nor are they "outside" of nature. For them, humans do not "transcend" nature. The anthropomorphism of their perspective is not a form of anthropocentrism. The Amazonian Indian sees a jaguar as a human and believes that the jaguar perceives itself as "human," with an inner life or "soul." The second fundamental point is the discussion of what is reasonably "enough" in terms of well-being, sustenance, and so on. This is a clear vote in favor of the integration of "limits" (reasonable sufficiency), against the view of some of the Dialogue participants that limits are unnecessary and stultifying.

10. Ignazio Musu contributes as an economist to the Dialogue; in his paper, he starts out by describing the difficulty of presenting oneself as an economist in a gathering of ecology-minded people. Economic growth and the relentless globalization of markets and supplies equals ecological degradation, in their minds; nor is that opinion without grounds (China is the example that comes to mind). For Musu, a responsible economist is one who endeavors to orient practical behavior (concrete actions) toward a model consistent with both economic growth and ecological balance. Global cooperation is needed, through national and international public policies. The Christian pattern of human beings as creatures in creation collaborating with the Creator has its usefulness; in China, government uses the model of "harmonious society" that should allow for economic expansion and for the conservation of environment.

11. The final paper is presented by two colleagues, Michael Shellenberger and Ted Nordhaus from the Breakthrough Institute (California). In "Modernization as Liberation Theology," the postenvironmentalists Shellenberger and Nordhaus question many assumptions of "eco-religion" and its "apocalyptic" narratives. The criticism is directed at the lack of consistency of élites who voice concern and even anxiety over excessive consumption, while increasing their actual consumerism. Eco-religion is attacked because of its orientation: it seeks to make believers turn away from the world, instead of embracing its imperfections and to invest it with the powers of human ingenuity. The rhetoric of downsizing and sacrifice is either hypocritical or a symptom of "technology anxiety." For Shellenberger and Nordhaus, only technology can save us from ecological disaster; and in order to enable a minimum of material wealth, education and health for the global population, the solution is not to reduce energy consumption but to replace the dependence on fossil fuels by "low-carbon" alternatives, for instance, nuclear power plants (see the Kaya formula that integrates the factors of global population, output of carbon dioxide emissions per capita, carbon intensity, and energy intensity). Indeed, to put it starkly: living in a hotter world is a "better problem" to deal with than living in a world without electricity. According to Shellenberger and Nordhaus, caution shows a lack of faith. The theology of modernization they champion unfolds a grand narrative of human evolution and development.

PART II: AFTERTHOUGHTS

This part consists of ten papers; a few participants were not able to contribute the additional paper. Nor are all the papers under the heading in fact "afterthoughts" in the sense of retrospective musings that continue to engage in the dialogue with the other participants; for different reasonings, the contributions of Bruno Latour, Ted Nordhaus, and Michael Shellenberger belong rather to the genre of statement, or manifesto: "My Manifesto for the Dialogue" and "Evolve: Modernization as the Road to Salvation."

Ariadne must have tied some knots on her thread through the Minotaur's labyrinth. On the thread that winds through the "Afterthoughts" I would like to tie the following knots, which are recurrent themes around which the participants exchanged controversy and common insights and proposals. The first of these knots is the first-person singular. The second "knot" is methodological: which discourse is most likely to "mobilize"? And what are the sources of our concepts and frameworks? The third and final knot is the problematical place of human beings in the universe: are human beings God's creatures, and/or are we "godlike," or even "god" (*demiurgos*)?

1. Thanks to the obstinate proddings of Pasquale Gagliardi, the participants who were able to offer "Afterthoughts" are impelled to say "I."

Simon Schaffer: "I learnt distressingly little during the Dialogue of how religious passions might ever mobilize a considered and judicious response to ecological crisis" (p. 230).

Izabela Jurasz: "I believe the real question is not about the possibility of mobilization, but the role that the religions already play in the ecological crisis [...] religious arguments can emerge on all sides" (p. 233).

Ignazio Musu: "We can reconcile the duty of preserving the gift of creation when transforming it. An attempt at this kind of reconciliation was made during the Dialogue, although I am in no position to judge whether it was successful" (p. 238).

Andrea Vicini's starts out with "I Have a Dream," the visionary speech (sermon really) given in 1963 by Martin Luther King. The memory of King's prophetic stance compels Vicini to formulate, also in the first-person singular: "It is this positive and inspiring dream in a just and sustainable future that led me to accept the invitation to our Dialogue" (p. 253).

Eric Geoffroy: "I would say that only a spiritual revolution, personal and collective, would seem to be able to meet the challenge" (p. 269). More specifically, Eric Geoffroy writes: "I'm increasingly convinced, at the end of this dialogue but also after wider reflections about what is going on in the world, that the crisis is global. It has various symptoms: cosmic, climatic, ecological, and—on a human scale—psychological, moral, and religious" (p. 270).

Elizabeth Theokritoff: "I was struck by the anthropocentric, or more accurately anthropomonist, tone that prevailed in much of our discussions (p. 274); I am very wary of apparent attempts to conflate human technological transformation of our environment with divine-human transfiguration of all creation" (p. 279).

Anne Marie Reijnen: "More strenuous efforts are needed, probably, to muster equanimity toward other people than to assimilate new information; I submit the difficulty has to do with values more than facts" (p. 282). "Protecting nature or saving creation [...] sounds like an alternative. The dividing line, I believe, runs between autonomous self-reliance—the ideal of secular science and ethics—on the one hand, and theonomous self-reliance on the other hand" (p. 284).

2. Regarding the sources of our diverse *Weltanschauungen*, for Izabela Jurasz, familiar with the early centuries of the Common Era, "Antiquity's leap toward the future—because it was a question of the future of the earth—was a significant challenge." But for her, "the example of Kepler gives us grounds to believe that the answers to ecological and religious questions can equally be found in Plato as in studies by climatologists and theologians" (p. 234).

In Andrea Vicini's contribution to the Afterthoughts, the legacy of Aristotle is alive and well: human beings are called to be what they are indeed by nature: "virtuous." This is key to the contemporary commitment to "sustainability." It is both feasible and desirable because of virtue. Virtuous personal and social behavior is aimed at achieving common interest, public interest, and common good: "greater global justice [...] for all citizens and for all creatures" (p. 255). Roman Catholic social teaching is well equipped for the task of searching "for better virtuous ways of articulating concretely our responsibility for creation by showing a greater respect for all living creatures" (p. 257).

Simon Schaffer argues that "making worlds" (the arts and the realm of *technè*) and the endeavor to know the world (scholars of all kinds) are not to be confused, but should be viewed as connected. "It is [...] disabling to imagine a nature entirely independent of, and unresponsive to, human purposes." In other words, "In the anthropocene epoch, with human and geological agencies so closely intertwined, it seems even harder than ever to disentangle a world that is *made* from a world that is *known*" (p. 231). This shows the lasting appeal of the image of a divine "watchmaker," who made things and knows them.

For Ignazio Musu, the dichotomy between praxis and theory weakens the effectiveness of Christian discourse. "Religious teaching continues only to be theoretical, consisting of preaching and declarations without any consistent action"; or, even worse: "what a religion's followers proclaim and what they do are often different" (p. 239).

In Michael Shellenberger's and Ted Nordhaus' account, theory and action of the ecological movement and of the Christian churches are consistent: "eco-theology"; the problem resides in the captivity of eco-theology to hypocrisy and a double morality: smug materialist consumerism at home and the need of limits for the "other," the developing countries. Their analysis has a distinctly sub-Marxian flavor: "Eco-theology, like all dominant religious narratives, serves the dominant forms of social and economic orga-nization in which it is embedded" (p. 244).

In Bruno Latour's "Manifesto," the readers are confronted with the need for a greater concord between religious eco-theology and the sciences. The Dialogue's aim is to "rejuvenate the links between science and technol-ogy, the various traditions of political ecology but also religion." Indeed, the epistemological boundaries between science and religion, "nature" and culture are antiquated, they are the leftovers from "first modernizations" (p. 249). The task for Christians is to reconnect with their cosmic concern and to renounce their obsession with morality. The same movement, but inverted, is demanded of sciences. The "second modern" scientists' *pen-sum* is to shift their values from "matters of fact" to "matters of concern" (p. 251).

Eric Geoffroy regrets that his is the only religious non-Christian contri-bution. The Koran does not validate a dichotomy between religion and sci-ence. *Shari'a* is both "religious" and "scientific"; it is cosmic law, not unlike *Dharma* (p. 269). The scholar of Islam also points to a similarity between Alfred North Whitehead's "panentheism" (quoted by Bruno Latour) and the Sufi Ibn 'Arabi (d. 1240), who sang the "Unity of being" (*wahdat al-wujûd*) (p. 271).

In the "Afterthoughts" of Anne Marie Reijnen, there is shift away from the centrality of words/hearing, to the realm of vision/seeing. The start-ing point has to be existential; it is important to retrieve the freshness of the first experience, the first time one saw the "blue orange," the first time one was confronted with the gruesome cross. As G. K. Chesterton said, "we see things fairly when we see them first." Therefore, she asks: "What happens to human beings when—out of the blue—they see their own habitat 'from outside'?" The cross is a message that can be under-stood only through words, the Word of God as it is witnessed in the Scriptures. The vision of Planet Earth needs no mediation. Traditionally, God has been apprehended through history, whereas today space becomes the "common ground." The cosmos becomes the "common place": the one place where we and all living beings can live, or where all meet their untimely death.

3. We use the word "knot" to identify a *topos* that interests several partici-pants in the Dialogue, and about which opinions diverge widely and maybe irreconcilably. The third of these "knots" is the question: "Like God(s)?": "*eritis sicut dii.*"

Ignazio Musu asks: is Man the Master of other creatures, or the steward of the gift?

In the contribution of Michael Shellenberger and Ted Nordhaus, the answer is "Yes! Like Gods. Better get used to it." Human beings are endowed with a godlike power to create their environment, to "re-create" the world. "What we call 'saving the Earth' will, in practice, require creating and re-creating it again and again for as long as humans inhabit it" (p. 241).

For Elizabeth Theokritoff, the answer is a resounding "No": only One saves, only One creates. Human beings are creatures; they are not God(s). Their technology can become a "counterfeit of salvation." She puts the question bluntly: "Does the ecological crisis remind us that we and all creation are ultimately dependent on God? Or does it reinforce a divide between man who 're-creates'—whatever that means—and a world that is to be molded according to his specifications?" (p. 274). A gulf separates the parts of the alternative: it is biblical faith versus (neo)pagan religiosity. However, the Jewish and Christian belief in God who "makes heaven and earth" is not a denial of human agency: for woman and man are called to be God's coworkers, in "synergy" with the Creator. "This synergistic approach, characteristic particularly of Eastern Christianity, facilitates a robust emphasis on human responsibility, without letting the weight of that responsibility become crushing" (p. 275). The question "Like God(s)?" receives yet another surprising answer. In a brilliant reversal where Christ is pivotal, the Orthodox theologian offers this startling perspective: "If our impact on the world reminds us that we are 'as gods,' this can be salutary. [...] To be 'as God' is to be on earth as one who serves, who lays down his life for all" (p. 277).

PART I

PROCEEDINGS OF THE DIALOGUE

1

Taking in the Real: Human Beings and the Earth

Opening Speech

Card. Angelo Scola
Patriarch of Venice

A Cue from Mahler

"*O Schönheit! O ewigen Liebens, Lebens trunk'ne Welt!*": "O beauty, O world drunk with eternal love and life!" These words that Mahler added to the text of the last movement of *Das Lied von der Erde* (1907–1909) arguably sum up the whole spirit of the work. They are fundamental concepts shaping the structure of the composition.

First, *beauty*. According to Prince Myshkin's celebrated claim in Dostoevsky's *The Idiot*, "Beauty will save the world."[1] But beauty, if separated from good and truth would, to use Dostoevsky's words again, this time pronounced by Dmitri Karamazov, be "terrible because it has not been fathomed and never can be fathomed, for God sets us nothing but riddles...The awful thing is that beauty is mysterious as well as terrible. God and the devil are fighting there and the battlefield is the heart of man."[2] And yet, as the great St. Augustine asks, significantly in *De musica*: "Tell me, I beg you, what else can one love if not beautiful things?"[3]

The second key concept in Mahler's phrase is the *world*, seen as the whole of reality. In this connection, his reference to *drunkenness* requires close scrutiny. It is not meant as an allusion to the "third eye of the poet" pointing the way to other worlds, which the so-called *poètes maudits* in late nineteenth-century Paris (Baudelaire, Verlaine, Rimbaud, Mallarmé, and others) sought by drinking absinthe. It is an opening up to fullness, overabundance, and even the *longing for*. This brings us to *love*, the power that "moves the sun and the stars,"[4] and often becomes solace in life. And lastly, *life* and *eternity*. Both because life is unquenchable thirst for eternity and because in every life there is something eternal.

Taking in the Real

Like all musical geniuses, Mahler alludes to an irreducible state of affairs. Reality speaks to man and man is able to take in reality. Indeed, there may well be an intimate correspondence between the two.

But where does the possibility of the relationship between man and the outside world come from? To come *ex abrupto* to the theme of this meeting: is this relationship the involvement—albeit at qualitatively different levels—of all beings in a single nature, or the relationship that both have with a Creator? Before attempting to answer this question, we must mention an important factor. Although the question concerning the relationship between man and the world is as old as humanity itself, today it has taken on an urgent new relevance. Unlike what happened up to the age of Kant, it now seems inconceivable that anthropological and ethical questions might come from cosmology. Considerations about the Earth no longer provide a picture in which man must find a place (anthropology); nor do they constitute an example to be followed[5] or to which man must or can refer in some way (ethics). Man now appears literally to be *im-mondo* ("not of the world" or "unclean" and excluded from the sacred). The Earth often appears only to be a kind of inconsequential ornament. People confidently go about their affairs but their affairs owe nothing at all to the cosmos. They are extraneous to it: "We no longer know in what way it is morally good that there are human beings in the world; and, for example, why it is good that they continue to be there. Is their existence worth the sacrifices that it costs? To the biosphere, to their parents, to themselves?"[6]

Precisely on these grounds, deciding what kind of relationship man has with the Earth is an urgent, crucial issue.

Man and the Earth

An initial suggestion as to what our position in the surrounding environment is comes from the Ecumenical Patriarch Bartholomew I of Constantinople: "It is a fact that the term 'environment' presupposes someone encompassed by it. The two realities involved include, on the one hand, human beings as the ones encompassed, and, on the other hand, the natural creation as the one that encompasses...we must clearly retain this distinction between nature as constituting the environment and humanity as encompassed by it."[7] Besides providing an essential initial description of the relationship between man and the environment, Bartholomew's remarks illustrate how this relationship belongs to the shared experience of life. Man experiences a living exchange with the created world and at the same time cannot avoid wondering about the meaning of being immersed in nature: where is that experience grounded?

In the Bible the environment in which man is created is represented by the figure of a garden (the Greek *parádeisos*), a place of beauty in which man's constituent relations—with self, with God, and all other living beings—are harmonious. Moreover, the "environment" itself has been created for man, who is called on to cultivate and care for it (Gen. 2:15). He is also given the task of naming the living creatures (Gen. 2:19).

Starting from theological thinking about creation, we realize how God's creative action is manifested not only in making the world exist, but also in making human beings free and therefore responsible for the whole of creation. The narrative of the Fall of man and woman is meant to signify that from the first instant of creation, man's freedom is at stake. We cannot think of man separately from his freedom. And the Earth exists for man so much that the Church identifies the root of the environmental issue in original sin. John Paul II described the issue in exquisitely anthropological terms:

> In his desire to have and to enjoy rather than to be and to grow, man consumes the resources of the earth and his own life in an excessive and disordered way. At the root of the senseless destruction of the natural environment lies an anthropological error, which unfortunately is widespread in our day. Man, who discovers his capacity to transform and, in a certain sense, create the world through his own work, forgets that this is always based on God's prior and original gift of the things that are. Man thinks that he can make arbitrary use of the Earth, subjecting it without restraint to his will, as though the Earth did not have its own requisites and a prior God given purpose, which man can indeed develop but must not betray. Instead of carrying out his role as a co-operator with God in the work of the creation, man sets himself up in place of God and thus ends up provoking a rebellion on the part of nature, which is more tyrannized than governed by him.[8]

This is why, as the Revelation still teaches us, the man-environment relation must be seen from the point of view of Redemption.

Christ's resurrection ushers in a new stage in which the relationship between man and creation is set under the sign of birth or "labor," which is painful but positive because intended for the good in life. And this is above all anthropological labor, which affects however, as St. Paul points out, the whole of creation: "For creation awaits with eager expectation the revelation of the children of God; for creation was made subject to futility, not of its own accord but because of the one who subjected it, in hope that creation itself would be set free from slavery to corruption and share in the glorious freedom of the children of God. We know that all creation is groaning in labor pains even until now; and not only that, but we ourselves, who have the first-fruits of the Spirit, we also groan within ourselves as we wait for adoption, the redemption of our bodies. For in hope we were saved" (Rom. 8:19–24). In this way anthropological labor and cosmological labor are interlocked in the ineluctable eschatological perspective. Thus in the second coming—already initiated on the path of the human family—what is already complete in Christ will be completed in us and in the world through the resurrection of our mortal body in our true body, in the *new heavens and the new Earth*. According to the Christian point of view, in this light we can look at the first creation and the new creation not as two separate realities that succeed each other mechanically, but as two moments that reciprocally embrace each other. The second assumes the first and gives its full meaning. The first in itself would inevitably remain incomplete and not adequately intelligible. Moreover, the historic-salvific path develops according to a plan conceived "before the foundation of the world" (Eph. 1:4), which will be realized in "the fullness of times" (Eph. 1:10).

With the new creation, Christ is revealed as the Head of creation itself:[9] the foundation of Christ's caring for all men until his death and his resurrection for us lies in the creation of all men in Christ.[10]

With thus grasp the literal meaning of creation in Christianity as the primordial relationship between God and the human person in the world: *Why did God create man and the world when he has no need of them?* This question can be couched in the terms of modernity as: *Why is there being rather than nothingness?*

Creation is the gift that God makes of Himself. Through it, he freely brings into being and maintains creatures in life, who, although radically distinct from Him, bear His indelible mark.

TWO REDUCTIVE VERSIONS OF THE MAN-NATURE RELATIONSHIP

This vision of existence enables us to eschew two inadequate conceptions—inadequate because basically incapable of fully accounting for human experience—of the man-environment relationship.

On one hand, an extreme anthropocentrism, whereby man is the absolute master of the cosmos. We know that some ecological thinkers base this line of reasoning on the precedence that the Bible accords to man over the created world.[11] The argument comes from the first version of the Genesis narrative of creation, which takes the form of an order given to man: "Be fertile and multiply; fill the earth and subdue it. Have dominion over the fish of the sea, the birds of the air, and all the living things that move on the earth" (Gen. 1:28). Without entering into a detailed reply to this critique, we can simply refer to the "second narrative" of creation (Gen. 2:41–3:24), in which the biblical teaching is formulated as follows: "The Lord God then took the man and settled him in the garden of Eden, to cultivate and care for it" (Gen. 2:15). Here there are not only two protagonists in the man-creation relationship—the human community and creation—but three, given that the relationship originates with the Creator. This leads to a further consideration. If man cannot rise to be the omnipotent master of the cosmos, nor can he delude himself that he can save it from disaster only through his own efforts, even when resorting to the remarkable discoveries and applications of science and technology.

Moreover, this prevents us from naively accepting a biocentrism or ecocentrism that sets out to "eliminate the ontological and axiological difference between man and other living beings, since the biosphere is considered a biotic unity with undifferentiated value."[12] Accordingly, "man's superior responsibility can be eliminated in favor of an egalitarian consideration of the 'dignity' of all living beings."[13] But this view impoverishes both the value of man, who is ultimately denied the status of a free agent participating in the activities of the Creator, and the value of the earth, which is stripped of all meaning that is not its own pure conservation. In fact as Pope Benedict XVI writes: "Human salvation cannot come from nature alone, understood in a purely naturalistic sense."[14]

If the cosmos is reduced to nature in which we are absorbed, our relationship with it can at most be aesthetical, but not ethical (Kierkegaard). Nature, however, is not only "a set of 'things' but also of 'meanings,'"[15] through which human freedom is called on to realize its own original vocation in the search for the face of the Creator.

ENVIRONMENTAL CONFLICTS AS AN ANTHROPOLOGICAL ISSUE

After this brief survey of the Christian vision of the relationship between man and creation, we can ask—in line with the objectives of the organizers of this event—if and how this conception, and similarly those of the other great religions, can still effectively interact with a way of perceiving and tackling the current intense ecological conflicts.

It is obviously not up to me, nor am I competent in the field, even to attempt to answer the question that will be discussed by the experts during the Dialogue. It may roughly be framed in the following terms: are religions, as demonstrated by their influence in other fields in the past, able to mobilize energies to contribute to a thoroughgoing ecological conversion? This would require a kind of *radical eschatology*, as Latour argues,[16] that is a long slow change affecting many areas of life referring to an enormous quantity of details and, most importantly, dependent on an infinite number of actions which, and this is the difficult part, demand that billions of people change their outlook. Can religious passions come to the aid of the low energy levels that seem to characterize the many ecological conflicts today?

This question contains a fairly overt invitation to frame in a radically new way the relationship between *eco-logy* and *theo-logy* in order to tackle openly the internal conflicts in these two worlds. I will only make a generic kind of suggestion.

I do not wish to go into the debate on the concept of nature. Almost everyone, in both the scientific and theological fields, now believes nature is doomed and considers this situation to be responsible for almost all the ills afflicting humanity. Personally, I believe that since *something given is always given to someone*, an ultimate ineffable element is ineliminable. And from Aristotle on, what has *fysis* been, if not this multiple, dynamic *actuality*?

But we must bear in mind, and especially as far as Christianity is concerned, that we cannot speak of nature other than in the terms of creation. And it is effective thinking on creation that paves the way to reconsidering the relationship between ecology and theology. Creation brings the *relationship* into the picture. Postmodern man is faced with a painful alternative. Having left behind the age of utopias and the pitch darkness it cast on the past century, postmodern anthropology has taken on a strongly Pascalian character. It is pursuing the meaningful wager of a radical alternative; does third-millennium man only wish to be the *experiment of himself* or does he wish to be a *self-in-relation?*[17] To face up to this challenge, anthropology must be dramatic. It must accept that the insuperable *one*, of which the self consists, is always present in a *twofold* way. I am one, that is why I can say "I," but I am always one of two: one of body-soul; one of man-woman; one of individual-community; and one of man-cosmos. Hence, otherness makes me an internal dimension of self, which on these grounds cannot exist other than in a relationship. It is the self that openly demonstrates this dramatic or polarized character. This is why the correct way of referring to the self is as the *self-in-relation*.

The interlocking of constituent polarities reveals the authentic relationship of creation as the permanent loving relationship of He who summons into being all reality (cf. Rom. 1:20) and continues to accompany it. According to the Jewish and Christian traditions, God made the relationship of love the reason for his compromise with the human family throughout its history. For the Jewish people and for Christians, he is *God with us*, and the *us* brings into play all the constituent polarities-relationships that I mentioned earlier. The ever polar relationship of self with oneself, with others, with the cosmos, and with God is the inevitable route by which we can say "I" in a humanly satisfactory way.

We inevitably see in this perspective the urgent task of inscribing the good relationship with creation in the intersecting circles of the other constituent relations.

I realize that what I am suggesting is too general not to run the risk of being obvious. But I feel it does show that there is a bridge between ecology and theology. And the more judicious scientists are also building this bridge today, having abandoned an ecologist vulgate based on a mythical return to a good and innocent nature. Baudelaire's exclamation—*Pan has come back!*—is empty. And we have even less reason for crediting Assmann when he describes Moses as an Egyptian. The way for the urgent, collaborative convergence between ecology and theology is to continue the logic of creation with love. This logic is scientific, religious, and political all in one. And consequently it is the logic of justice and of the complete development of humanity.

Religions can have something important to say on environmental issues when they are expressed through individual and social players willing to narrate the fullness of human experience and committed to putting forward valid arguments on its behalf.

Mahler himself bears witness to this when he says: "My heart is eternally devoured by a torment: my immense yearning for you."[18] Or when he feels he is prey to the questions that inexorably arise from experience common to all people: "Where have we come from? Where are we going to? Is it true, as

Schopenhauer says, that I really desired to live before being conceived? If I was created free, why does my personality imprison me? What is all this suffering for? How can cruelty and evil be the work of a merciful God? In the end, will death reveal the meaning of life to us?"[19]

As he was to tell his faithful disciple, Bruno Walter, on looking back on life when death already had a hand on his shoulder:

> There are many—too many—things that I could say about myself; I cannot even begin. I've suffered so much in these last eighteen months [after his daughter's death and his own illness] that I can barely tell you about them. How could I try and describe such a terrible crisis? I see everything in a completely new light; I have undergone such an incredible transformation that it wouldn't surprise me to find myself in a new body (like Faust in the last scene). I'm more eager than ever to live and I find "the habit of living" sweeter than ever.

He ends with a magnificent and particularly meaningful statement: "It is strange that when I hear music, even when I myself am conducting I find very precise replies to all my questions and everything is perfectly clear and obvious to me. Or rather, what I feel that I perceive with such clarity is that they are not questions at all."[20]

In short, after so many thoughts, desires, and struggles, Mahler finds true solace for his suffering in music—a real opening to the Mystery. The realm of music is very close to that of faith.

It is an opening inviting us to cross the whole of Creation.

NOTES

1. F. Dostoevskij, *L'idiota* (Milano: Feltrinelli, 2005[5]), 478.
2. F. Dostoevskij, *I fratelli Karamazov* (Torino: Einaudi, 1993), 144 (English trans. Constance Garnett).
3. St. Augustine, *De musica* VI, 13, 38.
4. Dante, *Paradiso* X XXIII,1 43.
5. Cf. R. Brague, *La saggezza del mondo. Storia dell'esperienza umana dell'universo* (Soveria Mannelli: Rubbettino, 2005).
6. Brague, *La saggezza del mondo*, 334.
7. Bartolomeo I, "A Sea at Risk, a Unity of Purpose," in *The Adriatic Sea. A Sea at Risk, a Unity of Purpose*, edited by N. Ascherson and A. Marshall (Athens: Religion, Science and the Environment, 2003).
8. John Paul II, *Centesimus Annus*, 37, www.vatican.va/holy_father/john_paul_ii /encyclicals/documents/hf_jp-ii_enc_01051991_centesimus-annus_it.html
9. Cf. H. U. von Balthasar, *Teodrammatica* 3 (Milano: Jaca Book, 1983), 33–39; 233–242.
10. Cf. H. U. von Balthasar, *Epilogo* (Milano: Jaca Book, 1994), 151–152. On the theological interpretation of Christ's salvific death, see G. Moioli, *Cristologia. Proposta sistematica* (Milano: Glossa, 1995[2]), 154–192; G. Biffi, "Soddisfazione vicaria o espiazione solidale?," in G. Biffi, *Tu solo il Signore. Saggi teologici d'altri tempi* (Casale Monferrato: Piemme, 1987), 42–67; H. U. von Balthasar,

Teodrammatica 4 (Milano: Jaca Book, 1986), 213–336; A. Scola, *Questioni di Antropologia Teologica* (Roma: Pul-Mursia, 1997²), 14–19.

11. Cf. G. Manzone, *Libertà cristiana e istituzioni* (Roma: Pul-Mursia, 1998), 140–141.

12. John Paul II, *Address to Conference on Environment and Health*, March 24, 1997, no. 5.

13. John Paul II, *Address to Conference...* For the debate on anthropocentrism, see S. Morandini, *Nel tempo dell'ecologia* (Bologna: EDB, 1999), 35–63; A. Auer, *Etica dell'ambiente* (Brescia: Queriniana, 1988), 201–220.

14. Benedict XVI, *Caritas in Veritate*, 48, www.vatican.va/holy_father/benedict _xvi/encyclicals/documents/hf_ben-xvi_enc_20090629_caritas-in -veritate_it.html

15. G. Crepaldi, "Il magistero della Chiesa e l'ecologia," in *Per il futuro della nostra terra. Prendersi cura della creazione*, edited by S. Morandini (Padova: Fondazione Lanza-Gregoriana Libreria Editrice, 2005).

16. B. Latour, "Si tu viens à perdre la Terre, à quoi te sert d'avoir sauvé ton âme?," *Revue-Théologicum.fr*, http://www.catho-theo.net/spip.php?article248

17. Cf. A. Scola, *Buone ragioni per la vita in comune. Religione, politica, economia* (Milano: Mondadori, 2010).

18. A. Liberman, *Gustav Mahler o el corazón abrumado* (Madrid: Altalena Editores, 1986),1 6.

19. B. Walter, *Gustav Mahler* (Roma: Editori Riuniti, 1981).

20. Ibid.

2

ECOLOGICAL CONFLICTS AND RELIGIOUS PASSIONS

A PATCHWORK OF BELIEFS*

"I think that a good way to start is to ask: how do we want to understand ecological conflicts? Primarily in terms of the narratives of society and history?" (MATTHEW ENGELKE)

"The idea of 'ecological conflict' or conflict in ecological matters depends on how we perceive our relation with the ecological world, a world of which we are part and in which we live as human beings. We are able to perceive a conflict in our relation with nature because of our special role in it, as living beings in a unique capacity to express conscious evaluation. The same is true for the idea of 'cooperation.' We often claim that cooperation is prevailing in natural life. But when we realize that this is not true, when we discover violence and even cruelty in natural life, independently of the action of human beings, then we are deeply disappointed. This is due to the fact that again the idea of 'cooperation' assumes a moral feature when linked to our special relation with nature as responsible living beings, when we make assessments according to some moral code. But what objectively happens in natural dynamics or evolution does not consider these assessments; it does not depend on them and it does not respond to them. We may ask, for instance, why there are conflicts, why there is what we define as innocent suffering in sentient beings in nature." (IGNAZIO MUSU)

"I'd like to explore how to allocate resources when we don't have the laws of nature—I mean 'pristine nature'—known by science to settle the dispute, I mean 'pristine nature' to settle disputes, and when we don't have the undisputable tenets of religion to settle disputes either." (BRUNO LATOUR)

"I assume that you would expect everyone to agree with when you say 'the sky is blue': it's not a problematic statement. But when you say that 'the cosmos is sacred,' that seems to be the core of the tension between a certain version of science and a certain version of religion, as it's reproduced in popular discourse that religion refuses to explain, that it is declarative." (MATTHEW ENGELKE)

"When we say that 'the cosmos is sacred' it is not something that we observe but a matter of belief, an affirmation of faith. It is something you can believe or disbelieve. We really should make our beliefs transparent, and that, indeed, methodologically comes first." (ANNE MARIE REIJNEN)

"If the central question is 'how the religious passions might be mobilized to address the ecological crisis and conflicts,' the starting point of our discussion should be: who is the sort of enemy you have to fight against, if there is a conflict between the development, on the one hand, and a sort of ecological concerns on the other." (TED NORDHAUS)

"What do we mean by 'environmentalism'? There may be a sort of environmentalism that is an ideology, but I am not sure how far I would accept that it is a quasi-religion. Certainly, some environmentalists are neo-pagans, which is definitely a religion—but most are not." (ELIZABETH THEOKRITOFF)

"Is the planet's overpopulation the problem, or is that not at all the problem? Can the planet support many more billions of human beings, or are there vast spaces which in fact need more population?" (ANNE MARIE REIJNEN)

"I feel that there is a total disconnect between the size of the claim that we are in an 'apocalyptic' situation—a religious term to be carefully analyzed—and the little emotion we feel." (BRUNO LATOUR)

"When we speak of protecting nature or saving creation, who is the subject? Is it the human species? Is it the Western civilization? Is it capitalism? One of my favorite quotes is by Fredric Jameson: 'Nowadays it seems easier to imagine the end of the world than to imagine the end of capitalism.'"(EDUARDO VIVEIROS DE CASTRO)

"There is no reason why someone can't be motivated by God and be a capitalist: these are not mutually exclusive understandings. We have to also consider where our passions for something within ourselves come into conflict, we can recognize conflicting passions vis-à-vis our long-term interests versus short-term interests." (MATTHEW ENGELKE)

"What is that makes us so resistant to the idea that we could solve the problem of global warming with producing power differently? Why do we want to suggest that there is something fundamentally wrong with human civilization? Why would we imagine that we have to get rid of capitalism in order to solve global warming?" (MICHAEL SHELLENBERGER)

"If we want to talk about the relationship between God and capitalism or between theology and sciences in general, the first thing we have to do is to abandon the traditional field of dialogue between science and religion, which is the field of metaphysics. According to Aristotle, meteorologists, astronomers and theologians deal with the same subject, that is they talk about so-called 'celestial' phenomena, situated 'beyond the physical world.' Therefore the discourse is 'metaphysical' in the original sense of the term, because celestial phenomena are *metà ton physikon*, beyond the world of *physis*, or beyond the world of nature." (IZABELA JURASZ)

"Post modernity has highlighted the limits of a metaphysical approach. Hence, today we need to find other ways to reflect on what goes beyond nature by keeping open the tension between anthropology and metaphysical discourse." (ANDREA VICINI)

"Do we believe that 'only a God can save us,' or do some of us want to say that only man can save us? If we are talking about agency, then this is a false alternative. I come from a tradition that believes strongly in synergy: only God can save us, but God will save us only if humans are prepared to co-operate with His work. When people talk about what is needed to save us from environmental catastrophe, there is more than one candidate for 'savior.' Perhaps these objects of hope are what from a Christian point of view would be seen as idols—or perhaps sometimes 'save' is used in the much more mundane sense of extricating us from a particular predicament." (ELIZABETH THEOKRITOFF)

"A passage in the Koran can perhaps describe the current situation, *sura* 99: the Earth is living, conscious, endowed with a memory and one 'Day'— some theologians claim that this is Judgment Day, but others disagree saying that it is the present and the current ecological crisis—'she' will bring up everything she has in her belly. When the Earth is finally fed up and can no longer put up with what humans impose on it, then she will tell the whole story of what man has inflicted on her." (ERIC GEOFFROY)

"Can it be a living being, the Earth? Is it a she or is it an it? Indeed, it is a she, and she will tell the story of all the bad things which we have done to her at the end." (BRUNO LATOUR)

"Is the Earth like a living being, capable of retribution? In biblical Scriptures, the earth is sometimes depicted as expressing God's justice: it flowers forth or withers, to express God's judgment. Such images, by the way, relativize the perceived anthropocentrism of the biblical narrative: the whole of creation, for example trees, fields and wild animals depends on God's mercy. In prophetical and apocalyptical discourse, human beings and their 'natural' world together are under the power of God's 'strong arm.' In the vision of Hildegard of Bingen (born in 1098), creation 'punishes' humanity for its excesses. This is an early example of the idea of a 'rebellion' of the Earth; the Earth acts not on its own, however, but as an instrument of divine justice. What about today's mentalities? In my opinion, to conceive of the Earth an autonomous agent of retribution who will 'retaliate' against human beings for their misdeeds, is the expression of a neo-pagan pathos that is not particularly helpful." (ANNE MARIE REIJNEN)

"We should distinguish ecological problems and ecological conflicts. Solving an ecological problem is very likely to entail ecological conflicts. This is why appealing only to technique in addressing an ecological problem may not be enough. The solution of the ozone layer depletion by means of replacing CFC (chlorofluorocarbon) was technically feasible and did not entail conflicts. With climate change things are much more difficult; not only for the moment the technical solution has not been found, and in this situation a lot of conflict dimensions emerge; maybe that when we will have succeeded in finding a breakthrough solution, conflicts will be reduced; this is not clear at all at the moment also because any transition path to a new energy system will bring in inevitable conflicts." (IGNAZIO MUSU)

"What I miss is the question of costs: almost all problems have a technological solution, but who is willing to foot the bill? We cannot talk about this without politics and without economics and without law; the legal dimension too is essential." (ANNE MARIE REIJNEN)

"What kind of God would foot the bill? Is it a Christian God? And then, what kind of Christian God, et cetera? Are we relying on a theological solution or on an anthropological solution? Not forgetting, obviously, that religion is a technology as well. Calling God for help is a technological measure. But, it is a question of theology too, because anthropology is in crisis just as much as any other idea, as it is no longer enough to deal with the problems we are living. And then metaphysics is back again into the scene, and it is alive and kicking. Somehow we can discuss metaphysics again without feeling guilty or particularly religious." (EDUARDO VIVEIROS DE CASTRO)

"The distinction between problems and conflicts reminds me of a typology I used with my students, when talking about organizational conflicts. Conflicts can be classified according to the means that can be used to cope with them. At the lowest level is what could be called the 'analytical conflict,' the conflict which comes from the ignorance or the uneven distribution of information among the actors involved. In this case, conflicts can be settled through analytical processes: it is enough to know more or to exchange information. If the conflict rises not because the actors have different information but because they have different interests, analytical processes are useless: actors have to negotiate. Finally, at the highest level, conflicts might spring from different value orientations: in this case, both the exchange of information and negotiation—whatever the amount of time and energy spent—will not bring you very far, they cannot easily settle the question. If the conflict is between not only 'different' but 'opposed' cultural values, there is no way of coping with it." (PASQUALE GAGLIARDI)

"Other criteria than the purely economic one must come in to address the extremely difficult task of transforming conflicts in some outcome which is cooperative. The question is whether religions have a role in this. There is evidence that in many cases religions do not help to transform conflict situations in cooperative situations. We have plenty of examples around us where religion forces and worsens conflicts. However ecology seems to be an area where dialogue between religions, not one religion acting alone, but religions talking together, may provide some values to sort out these conflicts." (IGNAZIO MUSU)

"If we are talking about where we place our hope of salvation—whatever that might mean to us—then this brings us to another whole area of ambiguities. It raises questions of what is meant by religion? What is meant by theology? In the way I would use the term, religion in the literal sense is an understanding, a worldview that involves a deity. And furthermore, religion has to do with not just what you believe intellectually but precisely with what you place your hope in, and what you worship. So this opens the door to a metaphorical use of the concept of religion. There might be other objects of hope and trust—an ideology, for instance, or indeed a technology—that could become the focus of pseudo- or quasi-religions. But it might be helpful to make clear when we are using 'religion' in this metaphorical sense." (ELIZABETH THEOKRITOFF)

Figure 2.1 *Grayson Perry, Map of Truths and Beliefs* (2011), wool and cotton tapestry, woven by Flanders Tapestries from files prepared at Factum Arte (290 × 690 cm, 114 1/8 × 271 5/8 in), courtesy the Artist and Victoria Miro, London.

NOTE

* The first session of the Dialogue was devoted to clarify the different meanings given by the participants to some crucial notions as (ecological) conflict, passion, religion, metaphysics, salvation, and the beliefs behind them. Actually, conceptualizations of the kind were given all along the Dialogue. The editors decided then to pick up the most pregnant of all these definitions, assembling them in a sort of preliminary "patchwork" or "map" of beliefs, so helping the reader to understand from the beginning the variety of positions and visions which were confronting each other during the Dialogue.

3

"Nature" or "Creation"?

Difficult Choices of the Church Fathers

Izabela Jurasz

"Protecting nature or saving creation"...Thus posed, the question is the cause of a certain perplexity for an assiduous reader of the Church Fathers. In fact, the worthy ancestors of Christian thought never really considered one or other of these two possibilities. That the idea of "protecting nature" is foreign to them does not surprise us too much. But that they have so little concern for "saving creation," seems to us to be very uncatholic!

Can we use texts of Christian Antiquity in the debate on the relationship between ecology and religion? There is much at stake, because—in a more or less clearly acknowledged way—all Christian theology inherits from the Patristic period: both with respect to the concepts to its sensitivity. In spite of the distance of two millennia, the primitive Church still remains a major reference, an undeniable authority. However, this reference requires a two-fold epistemological analysis. This twofold analysis will constitute the two parts of our reflection.

Given the distance that separates us from the Patristic period, it is essential to thoroughly study the concepts used by the Church Fathers, in order to "translate them" into our own language. In this process one often notes very many differences between what the Fathers really say and what we believe that they said or wanted to say...misunderstandings are discovered with not a few surprises! This is the first stage of the reflection that consists in highlighting the distance that separates us from the first Christians. The second stage, on the contrary, attempts to bring together the Patristic time and our own. Given the importance that is accorded to the Church Fathers—wrongly or rightly—how can they still inspire to us today is well worth thinking about. Admittedly, we speak differently and about different things, but there is perhaps a lesson to be learnt, a message to be deciphered?

THE HERMENEUTICAL DIFFERENCE

Deciphering the Concepts

The concepts in question were forged in the languages used by the early Christians; principally Greek and Latin, but also Hebrew and Syriac. We will thus start with a study of three terms—let us give them here in Greek: *ktisis* (creation), *poiesis* (production), and *phusis* (nature).

The first two terms are often interchangeable, but each has also its own specific use. One also associates with them derived terms, for example: *ktisma* (creature), *ktists* (creator), *poiema* (work, product), *poietes* (craftsman). There are also other synonyms, such as *demiourgos* (that which controls) or *plastids* (fashioner). Lastly, there are also terms employed as antonyms—and these antinomies are particularly revealing. It is not without significance to oppose production and creation, or the quality of sonship with that of the creature. This complexity deserves to be presented in some detail.

Ktisis

The first observation on the term *ktisis* relates to its eminently theological character. One cannot, in fact, speak about "creatures" or "creation," without having recognized beforehand a "creator." Admittedly, the idea of a creator is not the exclusive prerogative of Christians. It would be too long to enumerate here all the religions having a god or gods who made the entire universe or parts of it. On this point, Christianity finds itself the heir of both biblical thought and of ancient philosophy. Is there however a particularly Christian understanding of the term *ktisis*?

Christian thought initially is intent on identifying that which is Creator, much more than what is creature or creation. One important theme of reflection was to apply the title of Creator to the Trinity. Formulated in different ways we find this in Ireneus of Lyon, Athanasius of Alexandria, Augustine and Cyril of Alexandria.

However, the principal theological concern of the Christian authors is still elsewhere. In the wake of the Judaism, the first Christians apply the title of Creator to God the Father. But then, is Christ as God creator, or rather as man a creature? Paradoxically, to affirm that he is God and man appears less contradictory than to affirm that he is at the same time creator and creature, because the price to be paid would be to admit that he is a creator created, a creator who is not absolute and equal to the only Creative God. Against the Arians who spoke about Christ in terms of *ktisma* and *poiema*, the council of Nicea affirms that he is *gennema*, generated. Consequently, the quality of the creature will be opposed to that of the son.

This opposition is not indifferent for the Christians. Now salvation is to be "son in the Son" and not "creatures in the creature." Hence baptism, conversion and indeed resurrection at the end of time are all called "re-creation": a new creation or the second creation. Creation and creatures can be reconciled with the Creator only on the condition of being "recreated."

The concept of *ktisis* deals paradoxically with the Creator and not with creation itself. This concept reveals the complexity of the relationship between the Creator and his creation: the ontological difference that separates them can constantly be transformed into religious war. Without any doubt, this tension was aggravated by the christological paradigm, or rather by its misreading. This introduces an opposition between the created and the uncreated, between the human one and the divine one. To distinguish does not mean to oppose.

Poiesis

But worse is still to come: the sin of paganism, idolatry, consists in adoring the "creatures" in the place of the Creator. Here we come across a very close idea, that of *poiesis*: work. It is related to the *ktisis* in Arian discourse, where Christ is *ktisma kai poiema*, but it indicates a particular category of things created: a work carried out by a skillful artist according to the rules of his art.

The term *poiesis* remains, however, in withdrawal compared to *ktisis*, and that it seems for two reasons. The first is due to the close connection between *poiesis* and poetry. In fact, poetry is an ancestor of theology and the poets were the first theologians, severely reprimanded by the philosophers as "manufacturers" of false gods. Creation as a *poiesis* risks to become a make-believe creation, and a God *poietes* a craftsman who profits from human credulity. A fine line separates "art" from "artifice." The second reason for mistrust with regard to *poiesis* concerns the relationship with artistic creation. The craftsman creates from something: he organizes the matter that is at his disposal. Music and poetry are here an exception that nevertheless confirms the rule: God *poietes* is not the Master of the universe. Just like a sculptor is at the origin of the sculpture, but not of the marble. What is relationship then between sculptor and marble? If the first Christians spoke about creation "from nothing" to fight against the idea of the eternity of matter, their goal was not only to ensure the sovereignty of God over creation. Because preexistent matter became not only a concurrent divinity, but also was seen as a bad divinity, a principle of chaos and evil, a power opposed to God, the Father announced by Jesus Christ.

Thus, just when one thinks to have avoided the opposition between Creator and creature, one discovers another opposition, even more profound, that between God *poietes* and the "preexistent matter."

Phusis

What can this "preexistent matter" be? Sometimes it was indicated by the broad term of "nature," *phusis*. Just as today, in Antiquity nobody knew what exactly "nature" was. The first Christians found this term in the Greek philosophers and they used it with all the benefit of inaccuracy that it comprises.

The definitions of the term *phusis* by Christian authors first appear only around the sixth century. The lock of mutual agreement on the meaning of the term did not prevent its use. Quite the contrary! *Phusis* appears in the New Testament and quickly enters Christian theological language where it indicates completely different realities. Thus *phusis* can be synonymous with the term "substance" (*ousia*), but sometimes the two terms are carefully distinguished (cf. Cappadocians).

Phusis can indicate spiritual "creation" as well as material; it indicates the physical world and the laws that govern it. It can be also the "principle" (*arche*) of beings. There is a bit of irony in the fact that such a fuzzy concept as *phusis* can also mean all that is concrete, a reality, an obvious fact. To note a fact, the Greek man simply shouted: Phusei!—because what is obvious goes without definition, and the "naturalness" need not be explained. Curiously, the term *phusis* played a very important part in dogmatic theology, especially in christological and trinitarian reflection. Probably no other term divided Christendom as much—it is enough to think of the expression *mia phusis* and its applications for Christology because the *phusis* can be divine or human. It relates not only to the created universe, but also to Christ—according to post-Chalcedonian orthodoxy there would be a human *phusis* and another, divine—and also to the entire Trinity. According to the occurrences, the term *phusis* could indicate either the unity, or the individual character of each of the three divine Persons.

However, it should not be believed that the Christian authors are unaware of the complexity of the concept of *phusis*. Quite the contrary: the most important Fathers (Athanasius of Alexandria, Gregory de Nysse, Augustine) were not only conscious of the polysemy of the term *phusis*, but also of its use, differentiated to the point of contradiction. The difficulties raised by the three terms studied show well the hermeneutical distance that separates us from the Patristic time. They also constitute a series of warnings from which a fertile second reading of the ancient texts becomes possible.

COMPARISON

The Place of the Creator

The reflection on the range of the *ktisis* term highlights the need for thinking of That which we call "Creator" because creation is inseparable from the creator. To say that "God created the world" is largely insufficient, unless everybody not only believes in a same God, but has the same ideas about Him. Then other questions arise: how did He create the world? Out of what? Alone or with the help of intermediary workers?

The questions multiply ad infinitum and, without us being aware of it, we are building up a myth of creation. But there is an even greater risk, in my opinion—it is that these questions divert our attention from creation itself. The *ktisis* becomes a simple pretext for speaking about other things. The consequence is that Nature is reduced to a thin leaf of a fig tree, which renders the statues of Antiquity presentable to the eyes of the innocent ecologists. Then as long as the *ktisis* theme turns up somewhere in the writings of the Church Fathers, one can be at rest: for this is sufficient proof that environmental concern is quite obviously a Christian idea.

Should we then get rid of *ktisis* with the *ktistes*? Not necessarily. The same Church Fathers who ardently defended Christ "generated and not created," when they meditate on the book of the Genesis or the book of the Psalms, affirm that creation is good and that the created world makes it possible to discover the immensity of its Creator. But "Creator" is not the only name that says the relationship between God and creation. Another name is "Providence," *pronoia*, often found just beside *ktistes*. God as "Providence" exerts a benevolent presence in the midst of creation. His principal task is to fix the rules that govern the universe, but also to guarantee the freedom of these creatures.

It was in the name of God Providence that the first Christians defended free will against any form of determinism. Against the astral fatalism of the Chaldean Magi, Bardesanes, the third-century syriac philosopher, extended this freedom to the essential elements. They also—according to Bardesanes— have a certain degree of freedom, and for this reason they are subject to the judgment of God. Admittedly, the God Providence is thus also the God of judgment.

Defining "Creation"

The term "creation" has a double significance: an act (divine) and the result of this act. It is important to focus not only on the finished product, but on the very act of creation. Curiously, it is very rare that the Church Fathers support an idea of creation achieved in seven days—and that for many reasons. The symbolic system of the biblical numbers and/or the contradictions in the Genesis text is certainly one of them. But the primary reason for not insisting on the seven days is the eternity of God, who in these actions cannot be limited by time.

As a result they favor a deliberately allegorical reading of the account of Genesis, where the seven days are treated like seven stages of the creative act. Sometimes the Church Fathers do not even judge it necessary to speak about these seven days that so scandalize scientists, since in their manner of thinking about the act of creation they do not have a very great importance.

The Bible—followed in this by the Church Fathers—present two types of creative act: by the Word and in Wisdom because God created the world "by his Word" and "in his Wisdom." Creation by the Word is in the described in the book of Genesis: "God said and the world was." The act of creation is presented as being basically an act of communication. God creates when he speaks to his creation. The Word of Yahve—identified by the Christians with the Son of God—becomes the base (the principle!) on which all creation rests. And if the Son of God is also the Word (*logos*) of God, as a man Jesus of Nazareth is an answer of creation to the Creator.

But the act of creation is not only communication with, it is also an organization of the elements of the universe. This is Wisdom, another figure of the Old Testament subsequently identified with Christ, in spite of its feminine identification in the Bible. The sapiential model of creation is less known. Yet, the sapiential books speak about: Wisdom who builds a house, sets the table, and prepares a feast;—wisdom that God created "as a principle of his works"; and finally,—Wisdom who "lives among his/her children" "to teach them." The act of creation then appears as the work of the patient wisdom of the Creator. Wisdom builds the universe so as to live with her creatures. And she is always present there.

And why not translate the *logos* term as "reason" and speak about the divine reason that orders the universe? Admittedly, this solution represents the advantage of linking the two approaches: Word and Wisdom, but it involves taking a serious risk of impoverishing them. For a "reason" that orders remains all the same external to the ordered thing. Furthermore, one does not discuss established laws—one obeys them or not as the case may be. Room for debate would be quite restricted. But the pertinent question we must put to ourselves today is: do we still trust in reason, albeit divine reason?

Creature versus Nature

After the Creator and the act of creation comes the moment of the reflection on the creature. Creature or nature? If you like statistics, the occurrences of the term *phusis* are much more numerous in Christian literature than those of *ktisis* or *poiesis*. The concept of *phusis* is neither more nor less Christian than that of *ktisis*, but certainly more than *poiesis*. The awkward polysemy of the *phusis* term is not a contemporary invention either. If one reproaches the early Christians for incoherent use of the term "nature," it should be recognized that contemporary theologians are not more at ease with this term.

The concept of "nature" does make occasional appearances in the field of theology. One speaks about "divine nature" (in dogmatics) and about "human nature" (in moral theology), but seldom if ever of "nature" as an equivalent to "creation" or the "created world." This gap or vacuum partly explains the difficulty in the debate between religion and ecology.

What then does "nature" mean in the theological field? "To protect nature" or "to save creation" ... and why not "to save nature?"

Now, Christian literature—and not only Patristics—knows this expression well. Except that the "safety of nature" refers especially to human nature, "naturally" inclined toward sin. The ontological difference that separates the Creator from the creature is thus transformed into moral fault in the latter. There is no question of protecting nature in its present state! So, when applying the concept of nature to the created world, the idea of "saving it" implies its imperfection, its sinful condition. Its transformation then becomes a sacred duty. But before embarking on this crusade, we should reflect more deeply on the theological sense of the term "nature."

In conclusion, for religious arguments to be pertinent in the ecological debate today, it seems indispensable to reflect first of all on the theological connotation of the term "nature." We have shown that simply replacing "nature" by "creation" does not facilitate the dialogue between ecology and religion.

In the meantime, why not try to use the expression "to save nature"—both that of man and that of the world—emphasizing the "improvement" of that "nature!?"

DEBATE

BRUNO LATOUR

I have one first short question: I want to clarify the discussion between *ex nihilo* and *ex verbo*, because I'm not sure I understood if it makes a real difference in terms of concept and resource for the discussion.

IZABELA JURASZ

In Patristic thought we already have two kinds of creation *ex nihilo*, where *nihil* means both absolute nothingness and relative nothingness. It's very difficult to think absolute nothingness. It has a metaphysical sense, like the absence of being as such. On the other hand, relative nothingness is the absence of a form of being. In this sense, a thing can be called into existence from preexisting matter. And in fact the Church Fathers would have agreed with this preexisting matter. The question of the eternity of matter existed long before Greek philosophy and the Church Fathers saw no need to refute it. But at the time of the struggle against *Gnosis* the need arose to posit God as an absolute origin of everything—also and especially of matter—to ensure that matter was good because all of God's creatures are good. We must member that the Gnostics believed matter was the principle of evil, coeternal with God. This leads to a certain way of thinking about *creatio ex nihilo*, still remote from the current concept, which was developed by Thomas Aquinas in the thirteenth century.

For the Church Fathers, however, the eternity of matter was not an insurmountable problem. At most, they don't mention it much, because it is obvious that matter is eternal. For Theophilus of Antioch, for example, the fact that God organizes a primordial chaos, that he organizes this messy matter, is not theologically awkward. But, Tertullian, whom you know very well, was already arguing against the Gnostics and so he is forced to say there was absolutely nothing in the metaphysical sense before the divine intervention in the Creation.

SIMON SCHAFFER

I want to come back to the question of the watchmaker. In your presentation just now, you had a series of very quick maneuvers, and it would be very help-ful to go a little more slowly through them. First, you insinuate a distinc-tion between *artisan* and *artist,* and that's a distinction I find fantastically annoying, since part of the conflict around ecology and politics and engi-neering, precisely involves the claim that artisans are artists. So, do we need that distinction? And, is it helpful for you in a particular way? Secondly, you put the clock entirely on the side of the artisan that is, for an historian of the sciences, a very strange place for clocks and clockmakers to be, since for most of the period of the first Modernity the people who made clocks were called artists. Thirdly, because, as you know infinitely better than I do, one of the great tropes of Christian Natural Theology in the early Modern period, is "what kind of clock is this world? And what is the relationship between that clock, that is this world, and its maker?"

IZABELA JURASZ

I hope I have grasped your question on the difference between artist and artisan. In fact in Greek *poietes* means both. For the ancient Greeks, an artist like Phidias was an artisan. The distinctive feature of a *poietes* is exercising an activity according to rules. For Plato, people who make ships, people who make statues or invents myth are similar, because they all exercise a certain kind of activity according to the rules of their art, or their craft. They are all *poietes.* There is effectively a modern distinction that separates artist and artisan, but etymologically they have one thing in common—an activity conducted according to rules.

Thanks to the notion of *poietes,* we can go back to the unity and see that it is pretty close knit. Secondly, you speak about the clockmaker. Here my idea was simply to say that God is not a clockmaker because he remains inside his own creation. This is the idea of not separating the Creator from the cre-ation. The clockmaker is a familiar image—God creating the universe and then abandoning it to its own fate. Here, on the other hand, the idea is more that God will always remain within his creation, like an artist who in certain way continues to work. The Platonic image is the image of the poet. Plato says that a poetic creation exists at the same time as the poet pronouncing his discourse. The idea here would be of God as a musician. The musical work exists as long as the pianist goes on playing.

BRUNO LATOUR

The metaphor of the clock either means that the clock is detached from the maker and it goes its own way independently, or, on the contrary—that's what Izabela said, the clockmaker remains inside the clock he has made. God as an artist is continuously connected with God's work, in the same way as the musician cannot be separated from the music he is performing.

MICHAEL SHELLENBERGER

Artist/artisan: why modernity is responsible for this distinction? What is the argument?

SIMON SCHAFFER

Here the thought is, that ever since the *Renaissance* in Europe, there's been a fundamental distinction between artists and artisans and the various cognates of those two terms. And what's fascinating is that this distinction is not to be found, at all, in, for example, the Syriac and Greek of the early Church Fathers, where the word *poiesis*—we hear poet there—captures all that field. I have two kinds of problems with that story. To start, I have a problem with stories of the great fall that happened in the 1500s and 1600s, when tragically artisans were no longer considered to be artists, and fewer and fewer people participated in making the world. I do not think that is what happened in early modernity. And there is philological evidence for that in several languages: well into the 1800s the word artist is used, for example, in English and also in the Encyclopedia of Diderot and d'Alembert, to mean precisely clockmakers. Clockmakers are without exception *artist*.

This discourse is fundamentally connected with a term that we haven't used so far, which is the notion of genius. After the late eighteenth century, the work of the genius is always present in his/her products. It was for that reason that Kant and Schelling said that Newton was not a genius because you can follow his sums, whereas Beethoven is clearly a genius because you can't compose his music. Nowadays, to reclaim something genial in engineering and artfulness would be a very interesting move for this project.

ANDREA VICINI

Could you clarify your statement concerning "God is the artisan, the artistic creator and that concerns both good and evil?"

IZABELA JURASZ

Here what needs clarifying is rather the notion of good and evil. What is good? And what is evil? There are several answers: good as being, and evil as the absence of being, to quote Thomas Aquinas, for example. Without going into details, I could formulate my question as follows: in what way is God the author of good and evil? And how can man also be the author of good and evil? To answer this question, we must distinguish between good and evil in the metaphysical sense and in the moral sense. In fact, the domain of metaphysics is not our domain, just as it is not the domain of the Patristic writers. On the other hand, if we take good and evil in the moral sense—and morality only concerns man—then man may be at the origin of good and evil. He is not simply in a position to choose between one or the other, as if they were two objects outside him. But he is really at the origin of his actions, which are morally good or morally evil. I admit that when we speak of good and evil, the shift from metaphysics to morality is extremely easy and raises complex theological issues.

4

"Dame Nature Cares Nothing for Us"

Simon Schaffer

Creation, Production, and Survival

It has long been supposed that the power of the sciences is the impressive capacity to *shift* things from where they are and *activate them* when they get there. Yet in the case of political ecology it is argued that this mobilization no longer seems to work properly. The worry is that the sciences cannot now effectively shift, nor can they activate, agents of the ecological cause. There must, it is hoped, be a way to "grasp the world *otherwise.*"[1] This is the dilemma that invites reflection on *ecological conflicts and religious passions.* It would help to recognize just how entangled are these conflicts and passions. The connections are evident when, instead of conflicts about how a common world is seen by different groups, they instead concern the life of groups' very different worlds. Thus, at the end of 1994 the leading science journal *Nature* carried an editorial captioned "Apaches against stars." It referred to a letter from the Apache Survival Coalition protesting against the construction of an astronomical observatory on Mount Graham in Arizona as a threat to the sacred mountain and its many inhabitants, human and nonhuman. Amongst the observatory's users is the Vatican. Under the subheading "socio-religious concerns," the papal astronomers declared that "we do not expect the Apache nation to subject their divinities to the self-interest of a few any more than we would reduce the God of Abraham, Isaac, and Jacob, that is, God Our Father, to self-interested science."[2] Here ecological conflict's protagonists included Jesuit astronomers, Amerindian ecologists, stars, trees, squirrels, and gods, and the topic was, precisely, the passionate relation between divine entities and scientific self-interest.[3]

Jesuit astronomers on Mount Graham search for MACHOs (Massive Compact Halo Objects), heavy dark objects in the Galactic halo. Between 1992 and 1995 they met representatives of the San Carlos Apache nation to discuss threats to the *Gaans*, spiritual agents who "represented the elemental forces of the Universe" and would be disturbed by the observatory's work. "University officials, to say the least, were skeptical about the *Gaans* resting on Mt Graham." Anthropological connections between observers' cultures and the heavens' meanings were used to handle critical links between Jesuits and Apache. An Apache shaman deposed in 1992 that "in order to understand why there can not be any telescopes on Mount Graham you would have to understand our religion, language and culture, and you would have to know how to show respect...On this mountain is a great life-giving force. You have no knowledge of the place you are about to destroy." Initial response by a Jesuit ethnohistorian insisted on the late arrival of the Apache: "they were not a mountain-dwelling tribe...the sacredness is about as specific as references to the sky." One intractable aspect of the dilemma lay in this claim about aboriginality. As the eminent Harvard-trained member of the Lumbee Indian Tribe, law professor Robert Williams, has argued, the Mount Graham controversy seems to demonstrate inbuilt incommensurabilities between modes of existence. Yet, the Pope's servants cannot and do not easily perform the standard iconoclastic gesture of the naturalist. "We invite our Apache brothers and sisters," the Jesuit astronomers declared, "to join in finding the Spirit of the Mountains reflected in the brilliance of the night skies."[4]

Spiritual reflection is common in the maintenance and definition of such scientific sites during ecological crisis. One lynchpin of that science, thanks to the astonishing projects of the climate scientist Dave Keeling, has been an observatory used to monitor carbon dioxide levels in the atmosphere, built in 1956 atop the Hawaiian volcanic peak Mauna Loa. The long time series of rising carbon dioxide concentrations assembled at the observatory have become crucial matters of concern in the debate around anthropogenic climate change: so the precision with which the observatory's environment is managed turns into a decisive aspect of these disputes. And it is on the basis of this Hawaiian data that we are now told (in this case by the American economist Jeffrey Sachs) that "nature doesn't care how hard we tried; nature cares how high the parts per million mount."[5] So here nature has a subtle aspect, an entity whose performance *is valued* in parts per million, who *values* these measurements, yet who *devalues* any measures humans take. Questions of value—in all its senses—much affect the very possibility of making places where such measures are made. In January 2007, comparably fierce controversy broke out around rival claims that the nearby Hawaiian peak of Mauna Kea is the meeting place of the sky god Wakea and the Earth mother Papa, parents of the first Hawaiian ancestor, and that the great mountain is the very best place to build a USAF–funded $100 million telescope to track asteroids before they crash into Earth. This matter of concern has been debated between astronomers and members of the Hawaiian Culture Committee of the Mauna Kea Management Board. According to one of the astronomy professors, "for many years the astronomy community was through ignorance or arrogance insensitive to the sanctity of Mauna Kea to some Hawaiians. In their eagerness to build bigger and better telescopes, astronomers forgot that science is just one way of looking at the world and that we must be respectful of world views that differ from our own." Yet "science, too, is a culture, an ancient one," and attacks on the Mauna Kea observatories, so this astronomer urged, "ignore the kinship astronomers feel with the mountain as they explore the cosmos in what is ultimately a spiritual quest for them too."[6]

These passions of stars, mountains, spirits, telescopes, and gas particles recall conflict theses, stories in which scientific and religious communities are and always have been at war. As Stephen Toulmin has pointed out in his remarkable analysis of the seventeenth-century forging and the nineteenth- and twentieth-century dismantling of modern rationalism, "the alleged incompatibility of science and theology was a conflict *within* Modernity," not a long rearguard action against Modernity by diehard theologians.[7] In conjunctures of massive political, economic, and agrarian dislocation, these themes of credence and credulity were widely canvassed. This is certainly not a story of simple secularization, nor of the extinction of devout motives by self-interests. In political economy, the gods were also at work. Forged in the mid-eighteenth-century culture of enlightened political economy and social transformation, Adam Smith's science notoriously supposed that when individual agents single-mindedly pursued their own interests, the entire system would prosper under the guidance of what he began to call an *invisible hand*. In discussions of the possible relations between the order of religion and that of nature, it is worth recalling that the very phrase "invisible hand," which has played such a vital role in moderns' notions of agency and survival, was introduced by Smith first to describe what he saw as the primitive habit of attributing "the irregular events of nature" to "the *invisible hand* of Jupiter...Thus in the first ages of the world, the lowest and most pusillanimous superstition supplied the place of philosophy." Smith then reckoned that when enlightenment, seeking its escape from such idolatrous atavism, brought rational science to power, this "superstition" was systematically displaced by a very different invisible hand, that of private vice and public virtue. Faith in the "invisible hand" was thus no longer a sign of primitive fear but philosophic confidence. Yet for the Scottish professor, as for the participants in our Dialogue on protection and salvation, it is a real challenge to explain how and why such confidence could ever exert its power over locally entrenched interests. Smith held that it was the task of the sciences to allay wonder by showing how apparently discrete and startling phenomena were all connected in a single, wise, machinelike system. The sciences had the task of "representing the invisible chains which bind together these disjointed objects" and "introduce order into this chaos of jarring and discordant appearances."[8]

In Smithian cosmology, the source of natural order's benevolence was invisible. Compare the great early nineteenth-century radical journalist Richard Carlile, who whilst in jail for subversion composed a fierce polemic about famine, revolution, population, and political economy: "*Dame Nature cares nothing for us.* We must take care of ourselves in spite of her." "Though I have long and often used this word *Nature,*" the heretical Carlile continued, "I begin to see it to be one of those words which ignorance fashions to cover its nakedness."[9] It is no coincidence that Carlile entirely subscribed to the moderns' conflict thesis. Alongside his polemic about the senseless term *Nature,* he also composed a celebrated *Address to men of science* (1821): "we should view ourselves with the same feelings as we view the leaf which rises in the spring and falls in the autumn, and then serves no further purpose but to fertilize the earth for a fresh production," Carlile thundered, and added that "men of science have hitherto too much crouched to the established tyrannies of Kingcraft and Priestcraft." Even if, as Bruno Latour implies, "Christian churches were unable to digest the shock of science in the seventeenth century," the argument of radicals such as Carlile was just the opposite: that science had been entirely digested by the established church and had not yet liberated itself from that tyranny. Unusually for an English polemicist, Carlile even dared to damn both Francis Bacon and Isaac Newton as creatures of monarchy and slaves of religious superstition. So for this modern, even the greatest of modern heroes failed the test of liberation.[10]

The reason here to resuscitate the views of enlightened moderns such as Smith and Carlile is that their concerns directly interact with those of a project to reimagine the relation between creation, production, and survival. Protagonists of a range of schemes for the management of the Earth and of society, such writers were centrally concerned with the puzzles of attributing agency to inextricably linked social actors and natural powers. Latour reminds us that "religion never had much luck with *nature*" and judges as "a shameful episode" the enterprise of *natural theology,* in which the beauties of the cosmos were read as proofs of divine existence. These sad judgments accompany a history of loss. There is a resonance here with some other somewhat tragic versions of the moral and cosmological impact of the Scientific Revolution. Before the transformations of early modern European knowledge forms, there were surely a host of enterprises that sustained potent and polymorphous connections between microcosm and macrocosm. From then on, Latour argues, there were apparently but "two equally fatal exit strategies" available to a religion dealing with nature: in their mid-seventeenth-century versions, these were either Cartesian dualism ("a disembodied human soul") or Behmenist supernaturalism ("religion will try in vain to imitate scientific instruments").[11] Yet, it is also rightly stressed that fundamental versions of the modern dispensation were neither unchallenged nor hegemonic. Once again, philosophers such as Toulmin remind us of the paths not taken in the first modernity (which he identifies with Renaissance humanism's skepticism and antidualism) and the manifold ways in which the scaffolding of the moderns' key doctrines—nature has no history, matter is inert, causality is separate from reason—was dismantled. Toulmin includes, significantly, among the key moments of this disassembly the emergence of political ecology over the past half-century.[12]

In his compelling accounts of Amerindian perspectivism, Eduardo Viveiros de Castro gives some important motives to make such a move in the context of this debate on mobilization and ecology: for animism, he points out, "postulates a social character to the relation between humans and non-humans; the space between nature and society is itself social." Furthermore, he suggestively *identifies* the contrast between these cosmologies and versions of naturalism with the gift/commodity contrast.[13] In historical terms, therefore, it seems worthwhile to provide a nuanced account of the relation between the emergence of commodity fetishism and of modern naturalism. In what follows, a few remarks are offered about this kind of relation: first, with respect to the moderns' notion of *oeconomy*, then more specifically with contests around the *moral economy* that helped socialize the space between nature and society. A focus on the socialization of this space involves the familiar shift from a picture of social harmony to a picture of dissensus, in which rival speakers set out to legitimate their own account of nature in the interests of different models of the good society. Especially important in early modernity were differing accounts of nature as a place of agricultural labor. In the now classic account of the socialist historian E. P. Thompson and his colleagues, battles for social and natural power involved fights between a *moral economy* that celebrated use rights and custom in the name of a profoundly theological account of the sacred bounty of the soil and a *political economy* that transmuted such customs into cash-values and denied the godly origin of soil's fertility in order to justify the rule of the commodity.[14]

The site at which this political economy worked some of its most evident and fatal effects was within the Europeans' global imperium: any story of the social life of the spaces between society and nature must inevitably tell the terrible tale of the "late Victorian holocausts" recently analyzed by Mike Davis. His account links histories of climate—the atmospheric effects of el Niño—with histories of famine—the catastrophes of imperial agronomy in India and elsewhere. Smith's invisible hand dictated that "famine has never arisen from any other cause but the violence of government attempting by improper means to remedy the inconvenience of dearth." A powerful alliance of agronomists and astronomers then argued that natural laws (here represented as solar cycles) dictated life and death on Earth. It is common, perhaps, to imagine that the moderns were always interested in using their sciences to justify and then equip their active politico-social interventions. We need to recall, too, that the moderns also used their sciences to justify and then equip their fatal politico-social inertia. And this kind of fatalism is precisely the concern in our colloquium.[15] Late Victorian holocausts, and the accompanying models of ecology and economy in question in their course, are all too briefly evoked in closing this essay. But first, consider the genealogy of economy *as a social model of relations between human and nature.*

MORAL ECONOMY

It is appropriate to reflect on matters of mobilization, economy, and ecology in Venice, where four centuries ago the local mathematics professor Galileo Galilei and his allies devoted so much work to shifting and activating machines, guns, vessels, channels, banks, tubes, and ducts so that they could be subject to innovative forms of power over Society and Nature. The management of guns, as ballistics, and of waters, as hydraulics, allowed projection throughout the heavens and the Earth. There's been much to say about the relation between these kinds of political, economic, and hydraulic engineering, especially in tracing the roots of varieties of modern rationalism and authoritarianism. In his remarkable analysis of the relation between hydraulic engineering and the foundations of absolutist states, Karl Wittfogel pointed out half-a-century ago the coincidence of the great water programs of early modern Europe and the emergence of state-sponsored capitalism. Galileo's program and its near-contemporary versions in the Netherlands, France, and eastern Britain thus became modern archetypes of how the sciences acted remotely and mobilized the world. It was, for example, Galileo's unprecedented experiences of tidal rise and fall in the Venetian Lagoon and the Gulf, especially when travelling in barges carrying fresh water to the city from the *terra firma*, that first convinced him as early as 1595 that such water movements were the result of, and the best evidence for, the real mobility of this planet. Telescopes surely helped the new cosmology's cause; but barges and canals settled it.[16] What emerged here then was a new so-called science of waters, first among those geometrically expert Galilean scholars in central and northern Italy who would contrast their formal analysis of water flow with what they condemned as merely improvised tactics of state-employed engineers. Galileo sent the government a report on river management to establish his rights as "censor, perhaps among the most useful and necessary office," a representative of water flow, a judge of expert engineers, and a servant of the state.[17]

The great water projects of the first modernity, the construction of founts, channels, sewers, floating meadows, and drained lands, were pursued in the name of a kind of territorial *re-creation*, a return to a prelapsarian Eden on new terms set by new sciences. Francis Bacon, legislator of the modern project, said the new sciences could return humanity to Eden if their instruments were used well; nowhere was this more evident than in the reclamation of earth and the management of water.[18] One classical expression helped these moderns make sense of what was at stake in regaining such a Paradise: *economy* and its cognates. The model of state regulation that saw the prince as steward of the state's domestic order reckoned economy crucial to virtuous management of the political body. Long used solely for domestic steward-ship, we witness the sudden use of economy as public fiscal administration, social regulation, and frugal management only after the cataclysmic years of the mid-seventeenth century. This was not a strategy limited to hydraulic oeconomy—it was present in the entire program for improved agricultural economy that characterized early modern European societies. As histori-ans of natural history have taught us, naturalists won their power in society through showing their role in this oeconomy, now extended to include the means through which the deity managed His creation.[19]

The relation between expert knowledge and cunning art could therefore be understood, in part, by relating the magistrate's economy of territorial order in which experts would be hired to act, and the divine economy of Nature of which experts would be employed to make sense. The principles of Nature's economy could themselves be derived from Nature understood as a source of authority for human knowledge. Or they might be directed at overcoming Nature understood as a source of resistance to human purpose. Art showed what Nature truly was. The principal historian of enlighten-ment engineering, Antoine Picon, points out that "the engineers of Ponts-et-Chaussées were daily confronted by a hostile Nature against which they had to struggle to ensure the permanence of their works...Yet the engi-neers also saw there a principle of activity, a source and norm of behavior for humanity." This mattered especially in engineers' systems of long chains of reference designed to shift heavy-duty mobiles. "It seemed as if Nature turned its own forces against itself. Separating humans by means of Nature's contours, it nevertheless at the same time provided them with the means of reunification." Because Nature thus helped legitimate engineers' cunning reason while countering spatial engineering with recalcitrant resistances, the precise definition of Nature's economy was always also a definition of the rights of those who should be charged with its regulation and of those who would be directed and mobilized by the delegates of power. This enterprise was designed to move many worlds elsewhere. "Out of the economics of Eden," the environmental historian Richard Drayton argues, "had come an ideology of development."[20] As we will reiterate at the end of this paper, such an ideology had few Edenic consequences for its unfortunate subjects.

The conflicts between rival accounts of this economy hinged on conflicts about commodity forms. The existence of a *moral economy* reminds us that rationality cannot be seen as the prerogative of the moderns. In early mod-ern Europe, the agricultural market was governed by a moral order sanc-tioned by custom and Scripture. Agricultural labor transferred the divine product of God's earth from soil to table. The ultimate producer was God's action in earth. Divinity acted as fertility. Whoever intervened in this process by trading in the sacred territory between the fields and the home was guilty of blasphemy. A writer of 1718 condemned such traders as "a vagabond sort of people" who "have the mark of Cain, and like him wander from place to place, driving an interloping trade between the fair dealer and the honest consumer."[21] In riposte, political economists sought to place this language in the context of *superstition* and *vulgarity*. Smith's comparison between laws on the corn trade and on religion extended to his claim that the "popular fear of engrossing and forestalling may be compared to the popular terrors and suspicions of witchcraft." So rationally to repeal the corn laws would be comparable with the laudable act banning witchcraft persecutions. God's action in the soil and the existence of demons were equally implausible.[22] The capitalization of agricultural production, enclosure of common lands, and the disruption of traditional use rights were all contests over the plausible contents of nature and of nature's bounty as a commodity.[23]

Because of new demands for the legitimation of improving interests, systems where active principles circulated within the Earth and where the Earth was an object of sacred history were displaced by systems where the product of general agriculture circulated in an economy and the planet itself became an object of analysis. There was nothing natural in the drive for quantitative surveillance and the commodification of agronomy. Ideologues of the moneyed interest and of the improving landlords strove to naturalize new social regimes of calculation. In 1719, Daniel Defoe told the story of Robinson Crusoe's encounter with an apparently miraculous crop of barley on his island. Crusoe's initial response was to treat this bounty as a direct intervention by God in his affairs; then as a rationally explicable event, since he recalled emptying a bag of chicken feed on the spot where the grains were now growing; finally, he accepted the event as an example of the rational providence through which the world system was governed. Thus, Defoe made his hero recapitulate precisely the history of early modern theories of fertility—from the age of primitive superstition, when all events in the soil were attributed to the invisible hand of God, through skeptical atheism, when naturalists reckoned they could explain all such events by the action of natural causes alone, to the mature rationality of providential deism.[24]

The new agronomy was realized in the programs of the political economists and their intellectual allies. This was exactly when Adam Smith drew his telling calculation between "vulgar superstition" and prudent calculation, shifted the model of the "invisible hand" from primitive terror to rational confidence in self-interest, and satirized the moral economy in which "every object of nature...is supposed to act by the direction of some invisible and designing power. Does the earth pour forth an exuberant harvest? It is owing to the indulgence of Ceres." This "pusillanimous superstition," Smith reckoned, would disappear "when law has established order and security and subsistence ceases to be precarious." Rational philosophy would construct a system of laws that governed both the social and the natural order and no longer attribute the harvest to "the invisible hand" of the gods.[25] The moderns of the enlightenment at once applied these principles to agronomy. Smith's literary executor, responsible for the publication of the essay on the invisible hand of Jupiter, was his close friend the agronomist and earth theorist James Hutton. Hutton composed a vast agricultural essay of his own. He endorsed the virtues of the division of labor that his friend "Mr. Smith has so beautifully illustrated."[26]

Just as Smith argued in 1776 that the wise reason which governed the national economy was only to be made out by an impartial philosophical contemplation of the whole system, so Hutton reckoned that the systematic machine which governed Earth history was visible to the philosophical historian, not to those absorbed by the startling events of apparently convulsive nature. The way agronomists seized the soil was just the way the soil and the planet survived. "The foundation is laid for future continents, in order to support the system of this living world."[27] Nature, "considered as a machine," would always provide the agrarian basis for civil society, so rational philosophy's future was assured. And nature, this philosophy's topic, would never experience dramatic change, so its knowledge would be cumulative.[28] Hutton made rational agriculture the mark of humanity's transition to civil society. Each landlord was "like a God on earth" and such "a philosophic view" was "necessarily required in a person who shall, in acting deliberately or intentionally, control the course of nature ... such a thing is agriculture." This late eighteenth-century account of commodity agronomy depended on the identification of the real capital of scientific farming with the cultural capital of moral and social advance.[29]

Why Blame Nature?

In the catastrophic late eighteenth- and early nineteenth-century period of unemployment, high food prices, persecution, and World War, more radical experiments were proposed to change the agrarian world. This was the context of Richard Carlile's striking interventions in the debate on nature's capacities and human rights. Radical pastoralists set up collaborative schemes to take over land into production, and they were present in force in the cities; they were judged by the Committee of Secrecy and by police spies; their demands involved a reorganization of the social order in terms of the true principles of nature. As the working-class radicals' greatest historian puts it, "land always carries associations—of status, security, and rights—more profound than the value of its crop." This is how the principal radical agrarian campaigner made the point: "Tillage is a trade that never fails. But the usurpers of the soil shut out mankind from this best resource of nature as if they were the lords and creators of the earth and the rightful dispensers of life and death to the rest of their species."[30]

Thomas Spence, whose printed *Dream* of 1807 set out this graphic appeal to natural right in the earth and radical agrarian experiment, was a poor schoolteacher expelled from his local Philosophical Society for lecturing on these principles and campaigning with other reformers against enclosure. His immediate intellectual sources certainly included Scottish enlightenment agronomy. His manifestos were staged as a series of supplements and revisions of *Robinson Crusoe*. Under Spence's agrarian communism "trade will then be genuine, unforced and natural." At his funeral parade in 1814, his supporters constructed a huge statue of Justice to be carried through London streets with fertile soil samples laden into each of its balance pans. Spenceans told the crowd that "trade and commerce had been annihilated, but still the Earth was *by nature designed for the support of mankind*." Demonstrators marched behind a banner reading "Nature to feed the hungry." Moral economy taught that soil was possessed of unrestrained fertility, but Spenceans reckoned this alone was insufficient. The enemy, as Spence learnt from the Scots, was feudalism: landlords were no better than self-styled "landmakers," who stood between virtuous society and natural right. And the cure was a new science and a new constitution.[31]

At the same politically critical conjuncture of war and dearth, two Anglican priests published at length on the relation between creation's capacities, human welfare, and the puzzles of extermination. One of them, the country parson Thomas Robert Malthus, produced six editions of his *Essay on the Principle of Population* between 1798 and 1826 to demonstrate from the natural imbalance of population and resources the limits on social advance. The other, William Paley, one of the wealthiest churchmen in England, spent many pages of his *Natural Theology* (1802) explaining how Malthus's exterminism fitted into a divinely benevolent creation. "All superabundance supposes destruction or must destroy itself. Though there may be the appearance of failure in some of the details of Nature's works," the clergyman declared, "in her great purposes there never are."[32] Was nature or nature's god to blame? When Malthus attacked the Spenceans by name in the 1817 edition of his *Essay*, he was widely answered with the imagery of natural fertility and spade husbandry. Such debates gave currency to oft-cited remarks by agronomists such as the preeminent establishment chemist and agronomist Humphry Davy, who told his audience in 1813:

> The common laborer can never be enlightened by the general doctrines of philosophy...it is from the higher classes of the community, from the proprietors of the land, those who are fitted by their education to form enlightened plans, and by their fortunes to carry such plans into execution, it is from these that the principles of improvement must flow to the laboring classes of the community.

Against utopian visionaries for whom soil was naturally fertile and the inheritance of all, Davy lectured that "the vegetable kingdom" is not "a secure and inalterable inheritance" but "a doubtful and insecure possession, to be preserved only by labor and extended and perfected by ingenuity." The earth's natural fragility was exactly what justified its enlightened management by the agronomists and their market.[33] The economist allies of Smith and Malthus held that

> The notion of agriculture yielding a produce because nature concurs with human industry in the process of cultivation is *a mere fancy*. It is not from the produce, but from the price at which the produce is sold, and this price is got, *not because nature assists in the production*, but because it is the price which suits the consumption to the supply.

The mixture defined a kind of commodified "nature" that could deal with metropolitan precision and romantic pastoralism. As the historian of Enlightenment Roy Porter has told us, "by the close of the eighteenth century, Nature improved was becoming problematized, and…Nature became the new religion."[34]

Hybrid images of nature and of society are made up of visions like these. In the accounts of soil fertility, especially, we can see many salient features of later and contemporary versions of natural and social history. Following Davy, Justus von Liebig, and other masters of nineteenth-century agronomy, modernity does not see earthly nature as a sempiternal treasure house. Liebig famously wrote that "modern agriculture" was "a system of exhaustion" and often cited Smith's political economy to back his account of the farm cycle as an analogue of the circulation of industrial capital. The factory became the standard against which agricultural production was to be judged.[35] This model of nature's inevitable exhaustion played its most important role, of course, within the catastrophic alliance of colonial agronomies and sciences. Two discourses functioned simultaneously here: the imagery of energy and power engineering, of boundless global progress, of a kinship in the universal processes of life; and the imagery of decline, waste and death, of melancholy and impotence.

The twin stories of visionary engineering and apocalyptic waste remind us that there was an intimate relation between sciences that held that nature cared nothing for human welfare and that nature could be mastered to secure that welfare. Imperial astronomers pursued celestial signs that would allow them to predict the greatest colonial crises, the catastrophic famines that followed on the failure of Indian monsoons and of the subsequent rice and grain harvests. Even holocausts, so they judged, could be naturalized too. Successive government Famine Commissions recorded more than 20 major fatal crises of supply in the half-century after the 1857 Indian War of Independence. Political-economic orthodoxy, inherited from the first East India Company professor of political economy, Malthus himself, made state intervention in the grain market useless or worse. From their observatories based in the temperate hill-stations of the subcontinent, far above the sweltering plains, astronomers were convinced that "the same Indra and Váyu, the watery atmosphere and the wind, whom the Sanskrit race adored centuries before the commencement of our era, still decide each autumn the fate of the Indian people." By mastering solar cycles, Raj astronomers reckoned they could master what they judged an utterly natural phenomenon.[36] None of them conceded and many explicitly denied that "settled British rule" might in fact help cause these holocausts. As Mike Davis has demonstrated, free-market fatalism sat appallingly well with a search for famines' natural patterns. But the Parsi intellectual Dadabhai Naoroji, London professor and mathematician, angrily riposted that British administrators were responsible for the deaths of some 10 million Indians in the 1876–1879 famines: "why blame poor Nature when the fault lies at your own door?"[37]

This portentous question is linked with a moral history. Since they were sure devastating famines were the result entirely of monsoon failure rather than colonial economy, British scientists based in India worked hard from their mountain top observatories to collect data on the periodic fluctuations of the atmosphere and the Sun's energy. For two decades from 1904, these scientists established the existence of a vast global climate system, the Southern Oscillation, which linked together changes in the terrestrial atmosphere across the Indian and Pacific oceans. To return to the very start of our story, it was just the belated recognition of the effects of this Southern Oscillation that was decisive in Dave Keeling's claim in 1973 that in his Hawaiian data there was indeed evidence for a monotonic secular increase in atmospheric carbon dioxide due to fossil fuel consumption.[38] Thus, an imperial program launched to show that uncaring nature was responsible for catastrophe eventually helped demonstrate that human agency was culpable: from Raj atmospherics to the realities of the anthropocene. This is of course no defense of the vicious regime that first spawned that science. But such stories of the moral economies involved in agronomy, astrophysics, and climatology reveal many of the ways in which nature's seeming stoicism in the face of human suffering has long motivated admirable struggles for survival and liberation.

DEBATE

ANNE MARIE REIJNEN

Maybe the overarching theme of your presentation was the invisible hand, be it the providence or the forces of market, and the danger of fatalism and passivity, which is also a political tool. You started with the legal dispute, so we have economy, science, religion but also the law, the field of legal dispute about some contested territories. All these dimensions are always present, but it is helpful to make them explicit.

BRUNO LATOUR

Do I understand you saying that natural theology has never ceased?

SIMON SCHAFFER

I am not sure I have a grand narrative here, and one of the many reasons why I am so grateful to have been invited is to join a group of people who are much more able to define natural theology's scope. What struck me is the thought that natural theology is an embarrassing episode that you can simply brush away as a form of parenthesis because I take it to be one of the principal fields of the conflict that we are discussing here. It is a field of conflict mainly because the argument from design has had a complicated history in the enterprises, for instance, of Darwinism. Darwin is impossible without natural theology because he is convinced of adaptation.

The other question is: do these various forms of naturalization (which seem to me always to involve this kind of theological move, precisely a move when notions of the artistic and the artisanal quality of the world is appealed to and described) prompt activism, so re-creation, or passivity in the face of whatever crisis one faces?

These are resources which are used both to argue for comparatively massive modernization, of which English agriculture in the 1600s is a spectacular example, and on the other hand massively vicious passivity in the face of a Holocaust, to such an extent that Anne Marie Reijnen asks "What would the equivalent of the Shoah be, ecologically speaking?" One answer to that is Mike Davis's book.[39] Mike Davis shows us not just that the British administration of India between 1857 and 1914 killed millions of people, who should not have died at all, but that it was overdetermined by *El Niño*. It was the British astronomers naturalizing the famines who discovered for the first time the southern pressure oscillation that is the basic data set that eventually gives us the story about *El Niño*. So, one reason why I offer Davis's book as an example is because it shows the resonance between something that is environmental, undoubtedly, and something that we want to call political ecology, in this case vicious political ecology.

ANNE MARIE REIJNEN

So here we see science and pseudo-science as tools in the hands of a colonial empire. So again we're back to politics and ecology, science and superstition.

SIMON SCHAFFER

I would not want to use the phrase pseudo-science here. I want one to be aware that the astrophysicists who ran the observatories in India, who argued for the strong correlation between the behavior of the sun and the deaths of Indians, are not pseudo-scientists. They are scientists. And it's again no coincidence that the founder of *Nature*—the journal—is the same Norman Lockyer who invented this astrophysics. So *c'est le retour du refoulé*, one constantly is presented with this perverse spiritualization of a certain kind of science, and it's not a question of subverting true science.

ANDREA VICINI

Could you give us your own definition of theology of nature? The definition of natural theology from the point of view of an historian of science would be helpful and important.

SIMON SCHAFFER

To be very quick, in the period going roughly from 1660 to 1820, the First Modernity, in Britain natural theology meant two apparently contradictory things to its practitioners. On the one hand it meant the argument from design and was therefore extraordinarily controversial, since it did not seem plausible, especially to established Anglican theologians, that you could derive moral consequences from design. Indeed, I'm struck by a remark the Patriarch made in his allocution last night, that "Ever since Kant, we do not expect to derive morality from cosmology." Well, in England and Scotland it's both earlier and later than that. Lots of people would show you could not derive morality from cosmology long before Kant, and David Hume is a very good example. And on the other hand, however, the vast majority of their parishioners assumed that you can derive morality and the story of redemption from cosmology long after the publication of the *Critique of Pure Reason*, which doesn't seem to have much effect in English church pews.

The second claim was, that unless you are convinced that the world is designed, you cannot know it, which is the position Isaac Newton adopted. Unless you know this is a made world, it cannot be known at all, since one cannot know its principles, one cannot understand its architecture. The devices that enquirers build could not reveal how the world is, unless the world were a made thing. And that's also within the scope of natural theology. A very quick example of this: by far the most important physicist in nineteenth-century Britain, James Clerk Maxwell, in the 1860s and 1870s argued that the fundamental principle of natural theology is not adaptation but standardization. The reason we know the world is made is because there are uniform constants of nature, just as when we see a set of shoes all of the same dimension we are confident that they are manufactured articles. So the same must be true of atoms because they all vibrate at the same frequency and they are therefore created.

MATTHEW ENGELKE

From the anthropological point of view, the spokesperson of the Vatican observatory on Mount Graham is clearly missing the argument about autochthony and indigeneity, and this distinction is also of course what matters to Hawaiians. But what you have here is two competing discourses of rights, one based on being rooted in the soil, and one based on the right to know, that is science is for everyone, and it's never personal, it's for humanity, so there's a real tension there in the constitution of the understanding of a right.

SIMON SCHAFFER

If you want me to read out the text by Father Coyne,[40] I'll do it. "We do not expect the Apache nation to subject their divinities to the self-interest of a few anymore than we reduced the God of Abraham, Isaac and Jacob, that is God Our Father, to self-interested science." There is a fundamental contrast between the indigenous and the autochthonous, between living there and being of the soil. That's absolutely what's in play in Hawaii, because in the second line of this quote "some Hawaiians" covers a multitude of sins as you know extremely well, they're not the majority now in their own land. However, the second thought I think is more prickly because just as I will not let the phrase pseudo-science pass, I won't let the idea that a particular discourse or particular text is so obviously pragmatic that in a certain sense it's not serious. This is absolutely pragmatic writing by an astronomer and it's absolutely serious, I think. Or certainly it's as serious as religious discourse ought to be taken to be. I think it would be unfortunate if we reintroduced a kind of discursive dualism in which we find loads of evidence that the public discourse of the sciences mobilizes the resources of spirituality, and then we say: "Ah, but they're scientists, so they can't really be serious." That's a vicious circle that I don't want to get into.

BRUNO LATOUR

What is your indicator according to which you see that you move on the natural theology explicitly? In Newton it's explicit, in Maxwell it might not be so explicit. In the long run (late Modernism period) it was actually denied. So it does make a difference. I know you don't want to get into the periodization question, but after all you are an historian, you have to give us the answer to this question: what's the test that makes a distinction between people who deny there is a connection, like Dawkins, even though he is a simple case, and people who say there is a connection and that it's good. In other words, you cannot do as if the problem had always been there in the same way since you are also stressing radical differences.

SIMON SCHAFFER

I have a hypothesis, which is that the persistent announcement of the inevitable conflict between a certain form of science and other forms of belief—notably religious forms of belief—is often and perhaps always retrospective. That's to say, it is constantly pointing behind itself to the track record of the two enterprises. In other words, it's exactly as Izabela pointed our attention to, it's a discourse at the *deuxième niveau*. It's a discourse of the second level, in which one is constantly told that history shows us that we've always been at war. So my hypothesis is: rather than thinking of this as a period when there was a conflict, it's a series of political narrative maneuvers and techniques that reorganize the past of the sciences and other belief forms so as to show the permanent conflict. The two masterpieces of the conflict thesis by Draper and White, in the 1880s and 1890s in English, are both histories of the conflict between science and religion; and they go back to Roger Bacon or Galileo or Giordano Bruno or whatever. My hypothesis might be worth considering. Astronomers forgot that science is just one way of looking at the world. If someone says "back then," I would reply "When? When was that then?"

ANNE MARIE REIJNEN

The idea of Golden Age is very important. An imagined *Arché* is also an imagined *Eschaton*.

SIMON SCHAFFER

Exactly. The quickest way of saying it is that the conflict thesis is the bastard offspring of amnesia and nostalgia.

NOTES

1. Bruno Latour, "Will non-humans be saved? An argument in ecotheology," *Journal of the Royal Anthropological Institute* 15 (2009): 459–475, 466–468, 462,4 73.
2. "Apaches against stars," *Nature* 372 (December 15, 1994): 580; George V. Coyne, SJ, "Statement of the Vatican Observatory on the Mount Graham International Observatory and American Indian Peoples," http://vaticanobservatory .org/index.php/en/component/content/article/105
3. For such contrasts see Eduardo Viveiros de Castro, "Exchanging perspectives: the transformation of objects into subjects in Amerindian ontologies," *Common Knowledge* 10 (2004): 463–484, on p. 484.
4. See Robert A. Williams, "Large binocular telescopes, red squirrel piñatas and Apache sacred mountains: decolonizing environmental law in a multicultural world," *West Virginia Law Review* 96 (1994): 1133–1164, citations from pp. 1151 and 1162; John Welch, "White eyes' lies and the Battle for Dzil Nchaa Si'an," *American Indian Quarterly* 21 (1997): 75–109, citations from pp. 76 and 78; Coyne, "Statement of the Vatican Observatory." When the University of

Virginia, for example, considered joining the astronomy consortium at Mount Graham their delegation consisted of the provost, two astronomers and two anthropologists: "U.Va. officials to discuss Large Binocular telescope project," *University of Virginia News*, April 17, 2002.

5. Spencer Weart, *The Discovery of Global Warming*, new ed. (Cambridge, MA: Harvard University Press, 2008), 34–37; Charles D. Keeling, "Rewards and penalties of monitoring the Earth," *Annual Review of Energy and Environment* 23 (1998): 25–82. Jeffrey Sachs is quoted in Stephen Gardiner, *The Ethical Tragedy of Climate Change* (Oxford: Oxford University Press, 2011), 102.

6. "Three new telescopes approved for summit of Mauna Kea," *Honolulu Star Bulletin*, June 16, 2000; "Protest: no plan, no scopes," *Honolulu Advertiser*, January 26, 2007; Michael West, "Island voices: there's room for everybody on Mauna Kea," *Honolulu Advertiser*, February 17, 2003.

7. Stephen Toulmin, *Cosmopolis: The Hidden Agenda of Modernity* (Chicago: Chicago University Press, 1990), 144.

8. Adam Smith, *Inquiry into the Nature and Causes of the Wealth of Nations*, 2 vols. (London, 1776), 2, 119–120 and Adam Smith, *Essays on Philosophical Subjects*, edited by W. P. D. Wightman, J. C. Bryce, and I. S. Ross (Oxford: Oxford University Press, 1980), 45–46, 50. See Donald Winch, *Adam Smith's Politics* (Cambridge: Cambridge University Press, 1978).

9. Allen Davenport and Richard Carlile, "Agrarian equality," *The Republican* 10 (1824): 390–411, 404.

10. Richard Carlile, *An Address to Men of Science* (London, 1821), 6–9; Latour, "Will non-humans be saved?," 463.

11. Latour, "Will non-humans be saved?," 465.

12. Toulmin, *Cosmopolis*, 145–150, 163.

13. Viveiros de Castro, "Exchanging perspectives," 481.

14. E. P. Thompson, *Customs in Common* (London: Merlin, 1991), Chapter 4 ("The moral economy of the English crowd in the eighteenth century").

15. Mike Davis, *Late Victorian Holocausts: El Niño Famines and the Making of the Third World* (London: Verso, 2001), 31.

16. Karl Wittfogel, *Oriental Despotism: A Comparative Study of Total Power* (New Haven: Yale University Press, 1957), 32; Stillman Drake, *Galileo at Work* (Chicago: Chicago University Press, 1978), 37.

17. Drake, *Galileo at Work*, 320–329; C. S. Maffioli, *Out of Galileo: The Science of Waters 1628–1718* (Rotterdam: Erasmus, 1994), 51–65. For the attack on Sarpi see Frances Yates, "Paolo Sarpi's History of the Council of Trent," *Journal of the Warburg and Courtauld Institutes* 7 (1944): 123–143.

18. Paolo Rossi, *Francis Bacon: From Magic to Science* (Chicago: Chicago University Press, 1968), 129; Clarence Glacken, *Traces on the Rhodian Shore: Nature and Culture in Western Thought from Ancient Times to the End of the Eighteenth Century* (Berkeley: University of California, 1967), 471–475.

19. Lisbet Koerner, *Linnaeus: Nature and Nation* (Cambridge, MA: Harvard University Press, 1999), 100–104; Emma Spary, "Political, natural and bodily economies," in *Cultures of Natural History*, edited by N. Jardine, J. A. Secord, and E. C. Spary (Cambridge: Cambridge University Press, 1996), 178–196.

20. Antoine Picon, "L'Idée de Nature chez les Ingénieurs des Ponts et Chaussées," in *La Nature en Révolution 1750–1800*, edited by Andrée Corvol (Paris: L'Harmattan, 1993), 117–125, 117–119; Richard Drayton, *Nature's*

Government: Science, Imperial Britain and the Improvement of the World (New Haven: Yale University Press, 2000), 50–59.

21. Thompson, *Customs in Common*,2 08.
22. Smith, *Inquiry into the Nature*, vol. 2: 126, 119–120.
23. Joyce Appleby, *Economic Thought and Ideology in Seventeenth Century England* (Princeton: Princeton University Press, 1978), Chapter 9; E. P. Thompson, *Whigs and Hunters* (Harmondsworth: Penguin, 1977), Chapter 9.
24. Daniel Defoe, *The Life and Strange Surprizing Adventures of Robinson Crusoe* (London, 1719, reprinted Oxford: Clarendon, 1927), 88–90.
25. Smith, *Essays on Philosophical Subjects*, 49–50, 66.
26. Jean Jones, "James Hutton's agricultural researches and his life as a farmer," *Annals of Science* 42 (1985): 573–601.
27. James Hutton, "Theory of the Earth," *Transactions of the Royal Society of Edinburgh* 1 (1788), 209–304, 291 and *Abstract of a Dissertation Concerning the System of the Earth* (Edinburgh, 1785), 29.
28. Hutton, "Theory of the Earth," 291, 295 and *Theory of the Earth with Proofs and Illustrations* (Edinburgh, 1795), 2 vols., 2: 239. See R. Grant, "Hutton's Theory of the Earth," in *Images of the Earth*, edited by R. S. Porter and L. Jordanova (Chalfont St. Giles: BSHS, 1979), 23–38.
29. Hutton, *Investigations of the Principles of Knowledge* (1794), cited in Maureen McNeil, *Under the Banner of Science: Erasmus Darwin and His Age* (Manchester: Manchester University Press, 1987), 173; Sarah Wilmot, *"The Business of Improvement": Agriculture and Scientific Culture in Britain 1700–1870* (Reading: Historical Geography Research Group, 1990), 22–23, 40–41.
30. H. T. Dickinson (ed.), *The Political Works of Thomas Spence* (Newcastle: Avero, 1982), 119; E. P. Thompson, *The Making of the English Working Class* (Harmondsworth: Penguin, 1968), 254; Iain McCalman, *Radical Underworld: Prophets, Revolutionaries and Pornographers in London 1795–1840* (Cambridge: Cambridge University Press, 1988), 42–49; T. M. Parssinen, "Thomas Spence and the origins of English land nationalisation," *Journal of History of Ideas* 34 (1973):1 35–141.
31. Thompson, *Making of the English Working Class*, 176–179, 254; McCalman, *Radical Underworld*, 99; Roy Porter, *Enlightenment: Britain and the Creation of the Modern World* (London: Penguin Books, 2001), 460.
32. William Paley, *Natural Theology*, 12th ed. (London: Faulder, 1809), 479–480, and citation of Malthus at 505; Robert M. Young, *Darwin's Metaphor: Nature's Place in Victorian Culture* (Cambridge: Cambridge University Press, 1985), 27–31.
33. Malcolm Chase, *The People's Farm* (London: Breviary Stuff Publications, 2010), 134–140; Humphry Davy, *Elements of Agricultural Chemistry* (London: Longman, 1813), 24, 233; W. Krohn and W. Schaefer, "The origins and structure of agricultural chemistry," in *Perspectives on the Emergence of Scientific Disciplines*, edited by G. Lemaine, R. MacLeod, M. Mulkay, and P. Weingard (Paris: Mouton, 1976), 27–52.
34. Patricia James, *Population Malthus: His Life and Times* (London: Routledge, 1979), 278; Porter, *Enlightenment*,3 19.
35. Davy, *Elements of Agricultural Chemistry*, 233; Krohn and Schäfer, "The origins and structure of agricultural chemistry," 32, 36.
36. Norman Lockyer and W. W. Hunter, "Sunspots and famines," *Nineteenth Century* 2 (November 1877): 583–602.

37. Davis, *Late Victorian Holocausts*, 216–224, 323–326; D. Naoroji, *Poverty and Un-British Rule in India* (London, 1901), 216 is cited in ibid. 58.

38. Davis, *Late Victorian Holocausts*, 224–230; Richard Grove, *Ecology, Climate and Empire: Colonialism and Global Environmental History 1400–1940* (Cambridge: White Horse Press, 1997), 144–146; Keeling, "Rewards and penalties," 54–55.

39. Davis, *Late Victorian Holocausts*.

40. See note 30.

5

CREATION AND SALVATION

Bruno Latour

One of my specialties is in mapping scientific and technical controversies. "Controversies" is another word for conflicts. I am especially interested in ecological disputes such as, for instance, that around the red tuna, where you found fishermen, statistics about the reproduction of the stock of fish, Japanese taste for sushi, and so on and even President Sarkozy pledge to limit the disappearance of red tuna. It is a issue, typical of the many conflicts around resources. What is typical of those disputes is that there is, in addition to the usual economic interests, disputes about scientific data and technical solution. And to top it all there are conflicts around values and cosmologies. This is what drew me to political ecology and why I wrote *Politics of Nature*.

The reason I was so interested in preparing this meeting is because of the disconnect I mentioned between the horns of the Angels of the Apocalypse, on the one hand, and our laid back and rather *blasé* attitude toward what is supposed to be a total catastrophe, on the other. The discussions in Copenhagen in 2009 as we witnessed them prove the extreme difficulty of taking the measures that are at the scale of the challenge. I was greatly struck by Ted and Michael's remarks in their book *Breakthrough* that evangelical churches are full on Sunday while ecologists's gathering assemble only very few. Why is it, I asked myself, that ecology is not bringing in as much energy as the transformative power of religion has. But as a catholic, I was struck by another disconnect, this time between the tradition of the church, the very notion of Creation and Incarnation, the very idea of Salvation and those arguments about retreat, degrowth, and limits. So, again, *Breakthrough* struck a very important chord in me when I read it: here environmentalists were attacking the notion of limit in a very interesting way and I found it a very Catholic way. The goal is for me to break the notion of limits to growth, to detect what is so demobilizing in the notion of "we have reached the limits, now we should stop creating and innovating." How can creation be continued?

Now let me go through the several points of this argument. The first is an emblem or a simile of the position I am trying to argue. I want to retell Frankenstein's story (Figure 5.1). As you know, in Frankenstein story by Mary Shelley, Frankenstein is the name of the engineer not that of the creature, who is accusing himself of having sinned and he sinned, according to him at the end of the novel, because he innovated and he broke all the limits. His lesson is that he should not have invented it; he committed the sin of hubris. So, if you read the novel superficially, you end up saying, "Well, Victor Frankenstein is the apprentice sorcerer." But of course this is not what Shelley has meant. What Shelley meant was a very different story, a much more Christian story, which is that the sin, the real sin of Frankenstein, which is hidden by his blasphemous confession, is that he abandoned the creature. And there is a meeting, if you remember the novel, between Victor Frankenstein and the creature—which has no name—and they meet strangely enough on the Sea of Ice in the Alps. The creature asks the critical question, "Why did you abandon me?" Victor never makes anything out of this injunction of the creature. He never says, "I should not have abandoned you." He says, at the end of the novel: "Let me die and go back home, listen to my story and finally abstain from any sort of daring innovation." So, this story is important because it is connected with Simon's comments about genius and engineer. If technology is the god that may save us, which technology and what does it mean not to abandon the creation of our own making? God did not abandon his own creation—He sent his Son—while

Figure 5.1 Steel engraving for frontispiece to the revised edition of *Frankenstein* by Mary Shelley, published by Colburn and Bentley, London 1831. The novel was first published in 1818.

Source: http://en.wikipedia.org/wiki/File:Frontispiece_to_Frankenstein_1831.jpg

modernist engineers do. Frankenstein's story is a story where the engineers abandon their creatures and then, but then only, the creatures become bad. And that's actually what the creature says to Victor Frankenstein: "I was not born bad, but you abandoned me, and then I became wicked."

I think this is quite an important story for our discussion about what set of values do we mean when we say, "Now we will no longer abandon our creatures; we will not retreat from Creation, we will not abandon our creatures." And the whole question for me in this meeting is really: what do we mean by not abandoning creatures and what are the tests, what are the differences it makes, what are the shibboleths, the touchstone of this new set of values? That's the first point.

The second point has to do with modernism and the notion of a "change in trajectory" as many people use the term. I am not going to summarize the argument but I have argued elsewhere that "we have never been modern." As Simon said, we have always been doing natural theology as well. The idea of radical changes is largely a retrospective illusion on the front of modernization. Something else entirely has always been in the work. The other way to trace this situation is to use the word cosmology, a word that I am sure everyone agrees with here. And to describe it from the point of view of history of the sciences (which might be the point that is less well known around this table), let us describe cosmology and the never-ending natural theology described by historians of science as the distribution of qualities and properties between God, humanity, nonhumans, and matter. Of course, the question of periodization is very important, because it is not the same thing in the seventeenth century, in the eighteenth century, and now. However, if we have never been modern, it means we always have been attached to a certain definition of the distribution of the powers among all these entities.

Now, I am very worried by the fact of reducing the discussion of those "cosmograms" to the rather boring debate between Science and Religion for two reasons: that science was never doing what it was supposed to do since it never really abandoned natural theology; and that religion, at least in the Christian tradition, has always been doing something very different from establishing links between natural and supernatural. I tried to introduce a sort of different take by saying that it is almost the reverse: "When you talk about science look above—at the sky maybe, at the heaven—and when you look at religion, look down here." In fact, the traditional setup—where when you are supposed to talk about science, you are supposed to look down at matter, and when you talk about religion, you turn you eyes and you look above—is completely wrong. I try to move the conversation away from that scenography by saying that when we talk about science, we talk about access to the far away, to this long chain of intermediaries that I call "immutable mobiles," and when we talk about salvation and the saving power of the Christian *logos*, we talk about the near, *le prochain*. This means that already we have got completely out of the discussion between the natural and the supernatural, which is a sort of liberation.

The mediator is the third term: nature. Nature is out of the discussion because we have never been "in" nature anyway, or at least this is the redescription proposed by science studies. The Cartesian notion of *res extensa* has never extended everywhere. So, religious minds never have to fight a sort of idea of science that would have been extended everywhere, so there is no reason for Christian theologians to love-hate science, because science never delivered neither the thing to be hated, nor the thing to be loved. So, there is a sort of complete "comedy of errors" in all the discussions between science and religion.

I want now to introduce the word "creation." Izabela reminds us that there exist a lot of narratives in the Christian tradition that all do not go to the *ex nihilo* creation. I want to associate it rather with "creativity" (a word invented by Alfred North Whitehead), "creativity" in the sense of the innovative power of innovation of all entities. You can establish a lot of connections between all of these entities, connections that are uncontaminated, so to speak, by the "science versus religion" debate.

If we now consider the ecological crisis, I think it is offering a great occasion for rejuvenating Christian religion. Is it possible to move Christianity out of this anthropocentric and acosmic definition of itself? Maybe I put this in terms that are too apologetic, but since churches are empty, is it possible that, by harnessing the emotions, scale, and power of ecological crisis, we find way for Christian religion to abandon it is acosmism and to retrieve the sense of Creation instead of its obsession with nature and the supernatural? That was the sense of my paper quoted by the Patriarch: "*Si tu viens à perdre la Terre, à quoi te sers d'avoir sauvé ton âme?*," that is, the reverse of the way the Gospel's sentence has been understood until now. In fact, we know that the real tradition of the church has never been toward escapism.

In other words, I am interested in the conjunction between three issues. Issue one: can ecology renew what is meant by "Incarnation?" Issue two: thanks to the ecological crisis, could science come back to earth, literally, and jettison its natural theology? That is reaffirming that all the set of things that are supposed to be out of science—uncertainty, unintended consequences, politics—are back in. So, in the same way as Christian religion can be rejuvenated by the ecological crisis—because it will abandon it is acosmism—the same thing might be possible for the representation of science. Issue three: it is a great occasion for ecologists, thanks to the rejuvenation of Christian theology and the changes in the way science portrays itself, is it possible for ecologists to renew their strange forms of paganism.

I think that a very important topic for our discussion is that the prejudice against paganism in Christian religion might have been a missed target. The emblem here is not Frankenstein, it is Gaia. The intrusion of Gaia in the discussion might be an occasion for Christians to rethink what they meant by the entire history of idolatry, which probably was partly mistaken. And, of course, a great occasion for ecologists to rethink what they mean by not being anthropocentric. What does it mean for scientists, for ecologists, once freed from the spurious notion of "nature," to be not anthropocentric? A great argument for Gaia is that Gaia does not care for us. The whole argument of James Lovelock, if you remember, is in fact pretty well articulated: it is not a goddess, just a set of embedded and enmeshed retroactive loops, but if humans weigh too much on her, then it might just get rid of them. Gaia shrugs humans off, so to speak. My general argument is that Gaia summarizes all the interesting things that should happen to Christian religion, sciences, and ecology.

Imagine that in the middle of this table there is a sort of crucible: in this boiling pot we are throwing three old terms, religion, science, ecology, and out of it, after boiling them over, we get three totally different resources. To follow this operation of "transmutation," we have a lot of words and concepts to watch. "Apocalypse" is one of them. "Apocalypse" has clearly a very important meaning in theology, very elaborated and completely misread by apocalyptic discourse in ecology. It would be interesting to see the connection. Another one that we mentioned is "Innovation." Of course "Innovation" is a very powerful Christian as well as scientific and technical term, because "Creation" has its origin in "Innovation" and creativity is distributed everywhere. So, there are a lot of things to do with these words. "Asceticism" too, that could establish the link between Asceticism and degrowth, *décroissance*. But then, what about "Development?" This might be the best scenario since there are still 2 billion people without electricity. "Development" basically remains, but what about Asceticism? Is Asceticism the way out? Another word, which I thought would be discussed by the specialists of the Fathers, would be "Economy." It is a fascinating element that "Economy" is actually one of these words that can be actually shared by religion, science, and ecology. To quote Tertullian: "As better instructed through the Paraclete who is the leader into all truth, believe in one God, but subject to this dispensation that we call the economy."

I have got a lot of others words to watch. "Design," because "Design" is a term of engineering and the question of "what is good and bad design?" is a key issue to be shared by different disciplines. Another is "Principle of precaution," which is very important in European law. And "Principle of precaution" has a whole and very complex connection with "Limits," but also with the definition of the link between politics and science. Another one is "Sustainability," and "Eternity": what is the connection between the two?

So, there are many words to follow, and one of them to take out of the boiling pot is "Nature," this notion of nature as what we have been always "in." At last we are out of nature! Then science can be brought back down to earth and a great occasion is offered to rethink what ecology is about; because if ecology was confronted with a newly defined religion and a newly defined science, it would deeply modify the notion of limit. To give one example, and I will end on this: to have unintended consequences "attended to" is precisely the other word for "care." Finally, Gaia is seen by Christian religion as a demon or idolatry, but is it true? Because, in a way, Gaia is also a sort of test of the entire anthropocentric view, a way to render uncertain any position away from the human centered one, including that of God. We cannot avoid the meeting of Gaia with the Christian God.

DEBATE

ANNE MARIE REIJNEN

It seems to me you have showed us three unintended consequences of the crisis that are rather positive. It forces, it allows theology to rethink its own heritage and it also stimulates science to rethink its position versus religion. It also raises a question to ecology, of course, which is also a discipline that is historical, in that it has already history, and it is not unchanging. Would anyone like to comment right now on the double or triple meanings of the words that you mentioned at the end? Maybe not all have triple meanings, but I think they all have double meanings: "Apocalypse," "Innovation," "Asceticism," "Limits," the "*Shabbat*," "Gaia," "Economy," "Design" or "Pattern," and the "Principle of precaution." Would anyone like to come in here?

ELIZABETH THEOKRITOFF

Yes, could I ask for clarification of "Nature is no longer what we are in, we are now out of nature." "Nature" in what sense? I do not understand what this statement means.

BRUNO LATOUR

Well, to be "out of nature" is first a quote of course. And then it's the summary of a long argument pursued in the history of science, that we have never left natural theology. It is better to consider science anthropologically as a dispatcher, a distributor, a dispensator of capacities to many different entities some allowed to speak, some not, and so on. An anthropologist would find very different distributions in seventeenth-century England and twentieth-century France, of course, but would never find nature as *res extensa*, as what would be everywhere and what would generate a notion of matter to be then opposed to something like the "supernatural." So, in that sense, there is no big difficulty in putting nature off the table and using not necessarily creation, but creativity like Whitehead did. As long as nature is not put aside, we remain tied to the sad episodes of rational theology who swallowed the idealism of matter hook and sinker and accepted much too fast to believe that its job was to add the supernatural to nature or to dig in the "inner soul" some protected receptacle for religious values. A simple way to summarize the metaphysics of science in the history of science is to say "we have never been in nature."

Now, the second way to show that we have always been "out of nature" is the one of anthropology in general: most of the people on Earth have never lived "in nature" except some Europeans, according to the late seventeenth-century definition of modernity. Indians in the Amazon have never lived in nature; they have always lived in something entirely different made of lots of other entities that are distributed according to the patterns that Philippe Descola calls "animism" or "analogism."

A way to connect the two arguments above is to say that we have always lived in analogism; we never were naturalist, so to speak. We always mixed the values: politics, gods, creation, and so on, in the sixteenth, seventeenth, eighteenth, nineteenth, twentieth, twenty-first centuries. It is another way to say we have never been modern.

ANNE MARIE REIJNEN

So, nature is always an artifact?

BRUNO LATOUR

It is not only that "nature" is made, is that the notion of "nature" was never actually extent.

ANDREA VICINI

I would like to rethink what you indicated about nature, in order to achieve a correct understanding of nature in terms of human nature and of nature as such. What are we throwing away? What should we preserve? Personally, to preserve life on earth, a positive and well-defined understanding of nature might help us. We need a dynamic definition, continuously under revision and critique. It should result from a critical process. It should reveal our attempt to understand it better, without attaching to such an operative definition what we have described as a "bad understanding of natural theology." Which understanding of nature is helpful when we are concerned for the well-being of our planet? If I look at the theological tradition, I find both some insightful elements and others that need to be purified.

BRUNO LATOUR

Well, but the premise is that as soon as we stop having nature as the way to dictate a solution for conflict, and when it becomes dynamic and multiple or revisable, we enter into "multinaturalism." There is no conflict that can be simplified or settled by alluding to nature. In one word, you cannot use the protection of nature to do any thing. I agree that there is a set of very complicated issues here and that we have to be careful when we abandon "nature" because nature gives a lot of advantages to many, many positions. The question is: what position do we want to maintain? If you enter into controversies over ecology, if the simplifying, stabilizing verdict of nature is gone, then there seems to be no way to settle the disputes. What is the alternative?

When we talk about "multinaturalism" we are precisely in the conflicts that we are trying to resolve. So, what is the difference between good and bad nature? What is the good nature of tuna fish or red tuna fish? The tuna fish, the fisherman, and the legislator have completely different views of "nature." And this is precisely the question where the whole issue of ecological crisis is interesting because it forces us to use the word "multinaturalism," which is a new concept. Neither in the seventeenth century, nor in the twenty-first century does any one want to meet multinaturalism head-on. On the other hand, there exists no conflict that can be simplified by the appeal to one unified and agreed upon nature. So, when I say "out of nature" I don't mean "let's use outside nature as our undisputed and final referee," but what we have to invent when there is no longer any outside arbitration.

MATTHEW ENGELKE

I wanted to go back to Frankenstein and ask you if you think that what's being highlighted is the sin of abandonment, of being abandoned.

BRUNO LATOUR

The sin of abandoning, not of being abandoned.

MATTHEW ENGELKE

I wonder if this can be connected to the debates within theology and philosophy on free will and determinism. In other words, is there some kind of affinity between that debate and a parallel debate about abandonment and control? Because, of course, we can have a fear of abandonment, we can feel cheated by having been abandoned and yet the flip side to that is being smothered, is being controlled, particularly in a kind of parental role; a child is afraid of being abandoned by the parent, but the flip-side tension to that is not been given the space in which to become a human subject, to become yourself, to have that kind of freedom. So, I wonder if you can devote a few words to trying to relate these two debates within philosophy and theology about free will and determinism.

BRUNO LATOUR

I don't think we have that set of concepts in Shelley's novel. But we might have it in *The Tempest*.

SIMON SCHAFFER

If I can just amplify this: that oft-told story that is absolutely about the flip side of what Shelley is doing in Frankenstein, which is to settle the situation in a particularly colonial place, to separate the two creatures into one absolute body—Caliban—and one absolute spirit—Ariel. Ariel will be liberated and Caliban will not. Caliban rebels, Ariel will not; because Ariel will not rebel, Ariel becomes free; because Caliban rebels, Caliban will not become free. And at the end of the story, the redemptive act, which is exactly part of the implication of what Matthew is talking about, is the Victor Frankenstein figure—Prospero—who gives up the power to make anyone do anything, which is the exact point. So, it is not obvious what free will does in either of those stories.

EDUARDO VIVEIROS DE CASTRO

I had the impression that you had a double definition or double construct of what religion is. On one hand, you said that religion could be or should be a cosmology in the sense of having its acosmism expelled. On the other hand, you said that religion is about what is near at hand, is close at hand. That is to say: "*le lointain c'est pour les scientifiques, le prochain c'est pour les religieux*," "let's leave cosmology to scientists and let's keep the close, the inner self to religious people." How does that tie in with the recosmologization of Christian religion?

BRUNO LATOUR

When I speak of cosmology I don't allude to the usual way to compare religions with one another by looking at the ways in which they "think" of, for instance, God, matter, energy, and so on. This way of doing comparative religion, in my view, establishes a too simplistic way of comparing religions without ever to have to take the risk of speaking *religiously*, that is to try to explore in what way religion is not just another way of having "ideas" about the world, ideas that could then be compared too easily. Religion, as I take it here, is about a radical conversion and transformation of the world. So it is new in that sense, a new way to get into the cosmological question.

So, your question is very important. There is a tension, I agree, between the critique of Christian religion acosmism in recent time, and, on the other hand, my insistence on redirecting the religious gaze, so to speak, back to earth, back to *le prochain* instead of *le lointain*, best left to science. I think the answer to this tension is in the last chapter of *Process and Reality* when the question becomes that of "everlastingness." This is where speaking of Creation instead of Nature should make a difference. It is not coming back to the inner soul, or to morality or to anthropocentrism only. It is really trying to see if the transformative energy developed by a religion going now back to Earth instead of up in Heaven, back to Earth (rather with a capital E), allows the cosmological discussions to be themselves rejuvenated. On the one hand cosmology could look like some loose talk about ideas, some sort of an alternative Grand Narrative, on the other hand it is another way to "renew the earth," as in the prayer Christians sing to the Holy Spirit "who has renewed the face of the Earth." Does the transformative power of religion have a bearing on cosmology?

IGNAZIO MUSU

An important point made was that the ecological crisis is a challenge both for science and religion to rejuvenate. Science must understand that it cannot ignore uncertainty and unintended events; religion, particularly Christian religion, must rethink its central idea of contributing to re-creation. The reference to the true sin of Frankenstein may be helpful. This means that we should always care about the implications of our innovative activity. Two words are crucial in my opinion: "care" and "caution." "Care," because without care, without respect for other human beings and also for nonhumans, which eventually means "not abandoning creatures," we will not be able to succeed in dealing with the ecological crisis and ecological conflicts. And "caution": "caution" is very important when we address the point of technological innovation as an expression of modernization. To have caution does not mean to be against modernization; it simply means to be humble enough to understand that sometimes we may have ambiguous ways of using the possibility of modernization, in the sense that we should always ask ourselves if that specific innovation and the way it is used is appropriate to the task of "not abandoning creatures."

The importance of a positive answer to the challenge of the ecological crisis in terms of creativity and innovation, although with the provisos of care and caution, leads me to a very critical attitude toward the idea of *décroissance*, degrowth, as a way to deal with the ecological crisis. Sometimes I hear my fellow Catholics claiming that in order to respond to the ecological crisis we should fight against economic growth, we should accept "degrowth." I do not agree. I was, for example, happy to read in the last encyclical of Benedict XVI *"Caritas in veritate"* statements that appear critical on the idea of *décroissance*. Accepting this idea, in my opinion, is giving up the real challenge of the ecological crisis. Economic growth, as growth in consumption to meet basic needs, is an aspiration of the whole mankind. So, which right do we have of saying that we do want now to solve the problem of poor countries because we need *décroissance*? *Décroissance* in developed countries will not solve, it will aggravate the problem of poor countries. Address these issues with "caution" means that instead of claiming that we want *décroissance*, we should ask for a qualification of economic growth that does not give unconditional privilege to material goods, asking for a growth model that aims at improving the quality of life according to fairness criteria. But "caution" also means that we should not make the opposite mistake: that of aiming at a growth without limits. The sense of limit must be recovered, particularly when dealing with the opportunities of innovation.

TED NORDHAUS

In popular culture Frankenstein now refers to the monster and not the creator of the monster. Isn't Frankenstein now both the creature and the monster?

BRUNO LATOUR

All these expressions are there in the middle, in the boiling pot. Because *décroissance* can mean anything, from asceticism to the real sin of Frankenstein abandoning his creature instead of staying in his lab and bracing himself for new experiments. And we have to be very careful with "care" and "caution" because this could imply that we unfairly accuse the other engineers, who are doing things with care but are caring for other values. The same is true of "breaking the limits" because it is also an argument about creation, we too are creators. Nothing wrong with that. The question is what sort of creator we wish to be.

ANNE MARIE REIJNEN

It ties in with the excesses of the principle of precaution.

BRUNO LATOUR

And the principle of precaution itself is taken, by me at least, as an alternative to the paralysis of action under the power of science and one of the ways to rejuvenate politics.

ERIC GEOFFROY

In your text you make a very clear-cut alternative—at times I feel provocative in tone—between religion and nature: there would only appear to be either one or the other. I believe we can find a possible answer in the dialectic between transcendence and immanence. The Sufi Ibn 'Arabî (d. 1240), for example, tells us that we can only understand religion (Islam in this case), and reality in a whole integrative vision of the world, that combines especially the spiritual, psychological, and physiological levels. We must forge a complementary vision of transcendence and immanence. If the pole of transcendence is emphasized too much, we're deflected toward the fear of God, the separation between God and his creation, and theological and/or ritual fundamentalism; if we only see immanence, there is a danger that it is diluted into a confused monism. In your view, how can we elaborate this? If I've understood correctly, you say that we must incarnate religious or spiritual consciousness in matter, in the ground and especially in science. Is this not an answer to what you were saying earlier? Is it not a question of balance?

BRUNO LATOUR

A very quick answer is that in my metaphysics each of the three modes of existence (reproduction, reference, and religion) has its own definition of immanence and transcendence; so, immanence and transcendence cannot be used as terms of the metalanguage. That is the great trap that has caught dogmatic theology when it handles the sciences: it believed that it was a question of immanence and transcendence in the sense of the sciences (itself sorely distorted). Such a definition of transcendence has rationalized theology in the wrong way to enter into a strange hopeless and useless quest for the supernatural instead of questioning the natural to begin with. So, the whole argument requires to redefine immanence and transcendence in these three modes. The capacity of the organism to have an inner capacity for mutation is actually another definition for transcendence as well, which is absolutely unconnected to the idea of *res extensa*. So, immanence and transcendence are much too big to be brought in this discussion if you don't clarify what mode of existence you allude to. A balance between the two is for me fairly empty. They should not be used as terms of the metalanguage, science is just as transcendent in its own way as religion or organisms.

ERIC GEOFFROY

I agree with your idea that religion could be revitalized by the ecological crisis. Nonetheless, I believe that the crisis is not only ecological. I think it is global and takes on many aspects in the phenomenal world, such as fundamentalism, terrorism, and the various forms of moral crisis, depending on the cultures being considered (the ongoing Arab revolutions confirm the extent of the current upheavals).[1] Thanks to, or because of, the information revolution we know everything: hypocrisy, especially religious hypocrisy, is revealed to all. I believe it to be one of the signs of post-modernism. The ecological crisis is just one indicator among others.

ANNE MARIE REIJNEN

I would, at this time, like to read the question that was given to us from the audience, about *Le Contrat Naturel*, the classic by Michel Serres, *The Natural Contract*. The question is: is it possible to stipulate a contract with nature in order to cope with the conflict? What can a contract with nature, to cope with conflict, be? I take this to be a question to all of the speakers of today. It might be worthwhile to go back to the notion of religion as the contract between man and god. Would any of the speakers like to comment?

BRUNO LATOUR

I think that Michel Serres' book was very important in this conversation as precisely building a philosophical possibility, which has always been worked out by lawyers for many years. So, actually, there is a long conversation on a legal ground on this question of the rights, which has a lot of practical applications.

TED NORDHAUS

I wonder philosophically who we imagine were negotiating this contract. Who is on the other side of the negotiation? It is nature, as represented by whom?

ANNE MARIE REIJNEN

That was a basic objection to the book of Michel Serres. François Ost has talked about it in his book *La nature hors la loi*. Ost analyzes the impossibility of imagining an agent who is not human, because there is no deliberation, you cannot impute a fault to "nature," therefore there is no responsibility in the legal sense. In fact, it's a fiction. In this fiction, we are really negotiating this contract with ourselves.

BRUNO LATOUR

This old objection about the book—the fact that nature cannot be a contractor—shows that things have moved considerably in 30 years because at the time it was still nature as "the other" facing "the human."

NOTE

1. This was added later by the author, during the revision of his text in 2012.

6

HARMONIOUS COSMOS AND THE WORLD OF THE FALL: NATURAL AND COUNTERNATURAL IN THE ORTHODOX CHRISTIAN TRADITION

Elizabeth Theokritoff

I cannot begin without expressing my gratitude for the honor of this invitation and my joy at being here. It is rather daunting, trying to offer a contribution that could merit a week in such surroundings. Another, more serious reason for feeling somewhat daunted is that I feel I am swimming against the tide, in trying to make sense of some notions that seem to be regarded as unfashionable. One such is that of harmony in creation: how do we see this as being on some level a reality, not mere romanticism or wishful thinking? Or, on the other hand, the idea that the world we are living in is marked by a fall; that much of what we think of as "natural" is unimaginably different from the way the world was created to be and will ultimately become. Or again, the sort of linkage between the world and its Creator reflected in natural theology. In the Orthodox tradition, there is a theological understanding of nature rather different from natural theology as developed in the West. But it nonetheless prompts the question: How can we still believe today that the universe, the givenness of the universe, can be a source of guidance for the way we conduct ourselves within it?

I hope that my paper may also contribute another angle on some of the questions raised in other papers, and particularly the very crucial yet very difficult notion of *logos* in creation (i.e., word, reason—it can be translated in many different ways). Then there is the question of human greatness and what it consists in. What constitutes fulfilling our destiny as humans? A number of people have made the very important point that the environmental crisis should be seen as a call to humans to realize their greatness, not an excuse to diminish human nature; but everything depends here on what one means by greatness. I actually agree in one sense with a sentiment voiced with some alarm: that the destiny of man is to be grand, divine, and that people are spirits who strive to transcend. There is a sense in which Orthodox Christians would say that this is quite true; but we would maintain that man is not *only* spirit, and that achieving his "grand and divine destiny" requires, inter alia, recognizing that "enough is plenty" and that "environment is society." These are not opposites: the one is the way to the other. We achieve our grand and divine destiny by serving and not by dominating; and, since we are not only spirits but also embodied, we achieve it *with* all the rest of creation.

In this paper, I am not trying to present a general overview of the Orthodox Christian theology of creation. I would only note in passing that such an account would have many themes in common with Eric Geoffroy's presentation;[1] I am very struck by how much of the Eastern Christian world-view resurfaces in the cosmic theology of Islam, as he presents it. What I will attempt here is first to outline how the created order fits into the theological vision of Orthodox Christianity. Then I will explore at somewhat greater length the meanings of "nature" and "natural," and some of the paradoxes involved. How is it that Orthodox theology holds together such an exalted view of creation and man's place within it, and such a vivid awareness of the imperfections of our actual existence? I will refer particularly to the thought of St. Maximus the Confessor (seventh century), the theologian of creation par excellence. Then in conclusion, I will suggest some rather general ways in which the Orthodox vision might inform our interactions with the other creatures that make up our environment.

CREATION AND SALVATION

The starting point for Christian cosmology is not "man and nature" but "Creator and creation." It should be noted how this usage differs from the language currently in vogue among some green Christians, where "creation" seems to function as a pious synonym for "the environment." In the traditional understanding, by contrast, creation is everything that is not God—including, of course, man. Rather than living "in nature," we live *as part of* God's creation. In addition to the entire physical universe, creation also includes the angels; but since their destiny is not our concern, they can largely be left out of our discussion here.

Creation exists in relationship to God. There is an ancient and predominantly liturgical tradition of speaking of this relationship in terms of offering praise.[2] Angels exist to sing praise; nonrational created things offer "ontological praise" by being what they are and doing what they do. Humans are a unique hybrid: both material and spiritual, we are that part of the material world that consciously praises its Creator. This fundamental emphasis on man's role in creation as one of offering praise is strongly affirmed in Orthodox tradition by the importance ascribed to worship as a starting point for theology and Christian life. And also, indeed, by the character of worship itself, for example, the prominence of the Psalms, with their unparalleled "aesthetic appreciation of the inner life—and, indeed, inner voice—of nature *as a whole*."³ The liturgical day begins (at Vespers) with Psalm 104, a hymn of praise for all creation; Matins ends with Psalms 148–150, evoking the praise offered by all creation, including plenty of creatures inconvenient to humans. This forms the prelude to the essentially human offering of the Eucharistic Liturgy.

As creatures in the divine image, humans are endowed with freedom. This means that we can freely accept our unique role in offering praise to God with and for the rest of his creation; but we also have the capacity to reject that role. We are capable of choosing instead to see the rest of creation solely in relation to ourselves. If, however, we start to see usefulness to us as the sole function of other creatures, and to treat them accordingly, then we also risk obscuring the praise that they themselves offer. Mutual service and reciprocal needs are indeed fundamental to the way God has set up his creation; but to say that creation exists in relationship to God is to assert that the *usefulness* of created things, the ways in which they serve each other's needs, cannot be their primary raison d'être.

All things exist in relation to God; and, more specifically, in relation to Christ, in, through and for whom all things were created (cf. Col. 1:16). In the prevailing thinking of the Greek Church Fathers, articulated most fully by St. Maximus the Confessor, this relationship defines the nature of the created order. Everything that exists echoes the Creator Word (*Logos*, "rationality") through whom it was made; it is in him that the infinite variety of created things holds together as a coherent whole, a cosmos. Maximus even says at one point that the one divine Word "is" the many words of things, and the many words of things that exist are the one Word, who (also, on the other hand) exists in Himself without confusion, utterly beyond all creation.⁴ As a contemporary theologian sums up Maximus' thinking:

> According to Maximus, Christ the Creator-Logos has implanted in each created thing a characteristic *logos*, a "thought" or "word," which is the divine presence in that thing, God's intention for it, the inner essence of that thing, which makes it to be distinctively itself and at the same time draws it towards God.⁵

To affirm that all things are created "for Christ" is to affirm that all is created for *salvation*: the divine energy responsible for bringing the universe into being and sustaining it, says Maximus, is even now at work deifying the universe.⁶ So when I hear the words "saving creation," this is what springs to mind first of all: a cosmic process, also known as the divine economy. It is a process in which the Incarnation is central (which is why I find talk of a "Judaeo-Christian" view of creation quite unhelpful). It is daring but not, I think, inaccurate to speak of the entire history of creation as a "threefold embodiment" of the Word of God.⁷ The Word manifests himself first in the "words" or essential principles of the created order and then in the words of Scripture: and both these "embodiments" prepare for the appearance of the incarnate Christ himself, when the Creator enters into his own creation, taking up and literally making into his own flesh that which he has created. It is important for our discussion, however, to underline that the Incarnation is the first stage of a reciprocal process. God becomes what we are—not in order to confine us to a life of suffering and decay and "good" solutions that only bring new problems, but in order that we and our world may be transformed into what he is. By entering our earthly life, he inaugurates the state in which God will be all in all. Salvation is thus seen as the fulfillment of creation. For humans, salvation means fully manifesting the image and likeness of God (cf. Gen. 1:26); for the rest of material creation, it means being perfectly conformed to the "word," the inner principle of its creation by which it resonates in a unique way with the creative Word of God. This is a process that begins here on earth, but does not end here. We are thus dealing with a concept almost infinitely larger than preserving nature, preserving civilization, preserving human life on earth, even preserving the planet. We do not pretend to know what this "salvation" will look like. But we catch glimpses of it in the Resurrection and Transfiguration of Christ, when he is in the body yet transformed; he is no longer restricted by time and space. Even his clothing participates in his glory; man-made products can thus also be transformed. It is not apparent, however, that human products enjoy any privileged position in that process of transformation, in God's work of making all things new. We catch glimpses of "salvation" in the lives of people of great holiness (some of them our contemporaries), who live in peace with dangerous animals and whose prayers wield power over the elements. And we catch glimpses of it in the sacramental life of the Church: in the way that the matter of the world, water and bread and wine and oil, is given back to us as a gift of God's grace.

In some manner that we cannot pretend to imagine, just as man is destined for resurrection, so the entire universe is destined to be renewed and transformed into Christ. Man, the "hybrid creature," is essential to this process. As the material creature stamped with the divine image, he is the focus of the convergence of all things toward God. We thus believe that there could be no salvation for a "nature" that was by definition unrelated to man: there has to be some sense in which we "take up" all the rest of creation. On the other hand, this process does not necessarily have much in common with what we usually mean by "humanizing" our environment in the world as we know it. The idea of turning everything into a human product or amenity may be a matter of debate in regard to the earth, or even neighboring planets; but in relation to the entire universe, it is simply laughable. We "take things up"— and therefore potentially offer them up—not only when we use them or turn them into something of our own devising, but also when we study them or seek to understand them, or simply stand before them in awe.

If man is essential to the salvation of all creation, it is no less true that the "natural world" (nonhuman material creation) plays an essential role in guiding him to the fulfillment of his calling. For a start, it makes possible his life on earth—but it also does far more. This should not be surprising, since the Christian tradition has from the beginning seen the created order as God's self-revelation, a work that bears the hallmarks of its Maker. Perhaps the primary New Testament testimony to creation speaking of God is found in Romans 1:20, "Ever since the creation of the world his invisible nature, namely, his eternal power and deity, has been clearly perceived in the things that have been made." This picture relates to the idea that all things were made "very good" (Gen. 1:31), created "in wisdom" (Ps. 104:24): they form a *cosmos*, an arrangement of order and beauty. Evidence for this was traditionally seen in the way things work together, the harmony of apparent opposites so as to produce a certain stability and equilibrium, the regularity and order visible in the heavens and in the workings of the earth. Church Fathers such as St. Basil in his *Hexaemeron* also refer, for instance, to the way animals are adapted to their habitat and providentially equipped with what they need for survival, and the equitable way in which they seem to share resources (in contrast, the Fathers will often pointedly note, with human societies). This is the part that might easily be dismissed as a naïve and romanticizing view. But it certainly does not ignore all conflict; it clearly recognizes that conflicts of interest particularly between humans and other creatures are a feature of our nonparadisaical state. Rather, the Fathers are impressed by the fact that conflicts, tensions, and opposing forces—and constant change— nevertheless result in a world that functions and holds together, rather than in chaos and disintegration; and this leads them to the conclusion that God is in charge, that he is working his will even through events that on the narrower view might seem contrary to it. (By contrast, "nature red in tooth and claw" is the result of looking at the same data and focusing on the messy process rather than the outcome; and the natural conclusion is that nothing and no one is in charge.) Today, we obviously have a vastly more detailed knowledge of the processes by which things happen in the natural world, and the immensely complex pathways that have had to be navigated in order for the world we know to have come into being. It is by no means obvious, however, why this should make the notion of "wisdom" at work in creation any less cogent. Indeed, it is precisely the recurring hints that some sort of wisdom is at work that prompts theories such as the anthropic principle—which are not so easily dismissed as some of their "scientistic" detractors would wish.

The above passage of the Epistle to the Romans is fundamental to all Christian "natural theology"; but it would be a grave error simply to identify the vision in Romans with the later Western sense of a rational argument for a Creator based on the natural world rather than on "revelation." The great church historian Jaroslav Pelikan makes a distinction between "natural theology as apologetics" and "natural theology as presupposition"; it is the latter that is firmly rooted in Christian theology from the beginning. Indeed, it is hard to see how one can avoid some sort of "natural theology as presupposition" if one professes belief in a God who is "Maker of heaven and earth"— not to mention a belief that the Word of God "embodies himself" in some sense in the world that is made through him. Indeed, the early and Eastern way of looking at creation in relation to God might also be characterized as "natural revelation"—bearing in mind that, as Fr. Dumitru Staniloae states in the opening words of his *Dogmatic Theology*, "The Orthodox Church makes no separation between natural and supernatural revelation."[8] As Staniloae elaborates further on, "the inseparability of the two kinds of revelation and their content...would not be comprehensible were we to hold...that in natural revelation man is the only active agent." There can be no "separation of God from nature, a nature through which God speaks and works, or rather through which he speaks by working and works by speaking."[9] In Staniloae's sense, at least, "nature" is hardly something that the Christian tradition has "never had much luck with" (Latour).

As a revelation of God, creation is "universal" in that access to it is not limited by time, place, or culture; but, just as with revelation in scripture or in the person of Christ, recognizing it requires eyes to see and ears to hear. "The fool [who says in his heart: There is no God—Ps. 53:1] is foolish not at all because of deficient reasoning ability or defective inferences, but rather because of a hardly conceivable failure of spiritual vision."[10] The revelation through nature, like any other, needs to be appropriated by us in a way that is personal, even "interactive"; to switch from the "eyes" to the "ears" metaphor, the rationality, coherence, and interdependence of the created order is a "word" that is addressed to us and requires our response. There are principles here, says Maximus, that we can imitate in the way we live our lives, and in this way "reveal in ourselves all the majesty of the divine wisdom which is invisibly present in existent things."[11] Creation is not a source of propositions about God, but it is a source of vital insight into nature—in the theological sense of that term.

Nature and the Fall

"Nature" is a term of great importance in Orthodox theology. It has several meanings, but "non-human creation untouched by man" is not one of them. Creation as a whole may be termed "natures," that is all the various sorts of creature: when the Church Fathers speak of "the nature" in the singular in a concrete sense, they usually mean the human race. But underlying these concrete senses of the term is a more fundamental meaning: the "nature" of something (or of everything) is what it essentially is, according to God's purpose in creating it. Hence Maximus contrasts the "principle (*logos*) of nature" in each created thing, which is God-given and immutable, with the "mode of existence"—the way something lives out its nature in practice, which does not automatically accord with God's intention. A creature's mode of existence may be *according to* nature, *contrary to* nature or *above* nature. Human beings were created to move from their natural, creaturely state to a state beyond nature, becoming "partakers of the divine nature"— and taking visible things with them into a state of incorruption.[12] The tragedy of man is that he has chosen to go in the opposite direction, to embrace an approach to the world dominated by his physical senses—an approach natural to other animals, but contrary to man's true nature. Instead of seeing the world about him as a means of relationship with God, he sees it in relation to his own appetites. To use the symbolic language of Genesis, the tree in Eden is "good to eat" (Gen. 3:6). This results in a state of affairs in the world around us (what we commonly term "the natural world") *which is likewise contrary to nature*: it fails to correspond to its intended state. This is the drama described in the story of the Fall: in the traditional Christian understanding, this breakdown in man's relationship with God is fundamentally responsible for the difficulties of our life on earth, the dangers posed by the elements and predation among animals.

This is an enormous claim; and if one tries to fit the Fall into modern notions of history, there are obvious problems of temporal sequence given the overwhelming evidence for catastrophic events, death, and predation for millions of years before man came on the scene. But that is not the way the Church Fathers were thinking. Even in an age when no evidence of a prehuman past was so much as dreamed of, it seems that Fathers such as Gregory of Nyssa and Maximus did not envisage an actual historical period during which unfallen human beings were living on earth. The significance of paradise thus becomes more eschatological than protological: it is through our end, says Maximus, that we come to know our lost origins.[13] We might think in terms of the world being created in a certain way *in view of* man's fall, since God's knowledge of his creation is not restricted by time. The story of paradise then shows us what we and our world have the potential to become; the attentive reader of Genesis will note that the world is created *before* the Lord plants a garden, a "paradise," in one small corner of it. And the fall is our failure to fulfill the potential given to us to transform the world into paradise through our relationship to its Creator.

Despite the hermeneutical problems, the basic message of the creation-and-fall story is a very important one: that the world we live in is marked by two contrasting realities. On the one hand, it reflects the order, beauty, and goodness in and for which it is created; on the other, it exists in a "fallen" (counternatural) state in which survival and a degree of equilibrium are achieved through conflict, fear, suffering, and arduous toil: "In the world there is struggle" as St. Ephrem the Syrian says. It is tempting to speak loosely of the "fallen world" rather than the "world of the fall"; but it is not strictly accurate. All material creatures other than man serve God's will, as they were created to do; but their appointed task now also involves reminding man of his frailties and limitations. This sense that the world has two contrasting aspects is absolutely crucial to the Orthodox understanding of "saving creation" and of the human role in this process.

Quite apart from problems of how to interpret Genesis and a widespread feeling that the Fall story is too mythological to be of much practical interest, the idea of a world marked by a fall is not especially popular today. It is unpopular with many environmentalists because it is seen (wrongly) as denigrating nonhuman natures or suggesting that everything material is tainted, and (rightly) as signaling that no degree of conversion to green living is going to re-create an Edenic golden age of harmony. And it is unpopular with those who put their hope in technological progress and rising standards of living because it signals that hardships, instability, and choices between evils are unavoidable facts of life. This in turn suggests that the culture of consumption and exploitation of resources that has allowed unprecedented numbers of humans to enjoy unprecedented ease for the last few microseconds of the earth's history may not be able to continue for very long.

The idea that there is both a "natural" order of things which has to be (re)gained, and an accustomed order which is in fact counternatural and contingent on man's fall, actually helps resolve some of the dilemmas involved in the way we regard and treat our environment. We do not have to choose between "earth cursed" and "earth blessed": we can be quite realistic about the imperfections of the world and the struggles involved in living in it, while also revering it as a revelation and a faithful servant of our common Creator, marked forever by his incarnate presence within it. Nor do we have to choose between affirming man's dignity and recognizing his sin. Orthodox theology accepts that man has become the pollution of the earth, in a far more radical sense than any deep ecologist realizes; but this is attributed not by any means to man's nature, but to his counternatural mode of existence. The capacity of humans to degrade the earth is the shadow side of our calling to be the instrument, the "workshop" of its salvation. In an alternative metaphor, the Church Fathers like to speak of man in creation as a "king": think of the capacity of a corrupt ruler to turn a rich, fertile, and beautiful country into a hellhole of violence and misery.

GARMENTS OF SKIN AND NATURAL CONTEMPLATION

So the world, humanity included, presents us with two contrasting realities: the counternatural state which we mistake for natural, and the truly natural state which we may glimpse underlying it. This paradox is addressed at length by physicist and theologian Christopher Knight, in what is probably the most systematic attempt in English to explore the Orthodox theological understanding of the natural world: "We want to say both that this is God's world and that it is not yet *fully* that world as God intends it to be"; and again, "The world as we know it is both transparent to God's purposes and, at the same time, partially opaque to them."[14] Corresponding to the "not yet" and "already" are two aspects of our functioning within this world and interaction with the rest of creation: "garments of skin" and natural contemplation.

I use the term "garments of skin" (cf. Gen. 3:21) in the sense developed at length by the twentieth-century-theologian Panayiotis Nellas,[15] drawing on Patristic thought, especially Gregory of Nyssa and Maximus. It does *not* denote the body, which is seen as basic to the human make-up. The "garments of skin" added to man "like a second nature" denote our mortal and corruptible state, but also all those aspects of human existence that enable us to survive in this state: arts, skills and technology, learning and science, law, culture and politics, and so on. These things are at once consequences of our fallen state and a blessing: they provide us with a degree of hard-won "dominion" over the earth and its creatures that allows us to survive; they enable us to form reasonably stable and just societies, and even help us to restore the image of God's grace in the world about us. They serve as a blessing, however, only so long as they are recognized as *relative* goods. The "garments of skin" are ways of coping with the imperfections of the world, turning them to physical and spiritual advantage—not eliminating them. It is these very imperfections, and our constant failure to attain the good things we are searching for, that point us to something more perfect, to a destiny for all creation far beyond anything that our skills can contrive.[16]

Our ability to use, manipulate and organize the world, then, is God-given, even while the forms it takes reflect the world of the fall (so it is small wonder that similar capacities are found to a greater or lesser extent in other animals). But this does not imply an automatic divine seal of approval on whatever we might choose to do with those abilities. Again, we are faced with two possibilities. We can use our powers according to nature, so that we are working toward the fulfillment of God's purposes for his creation. Or we can use them in a counternatural way, which may however be highly effective in the short term, enabling us to outcompete other people and other creatures and thrive in a world marked by the Fall. Modern and future technologies—like all arts and technologies throughout human history—may take either of these forms. But how do we determine which is which?

A part of the answer, at least, takes us to our other mode of interacting with the natural world, which I have described rather loosely as "natural contemplation." This requires a capacity that belongs to our fundamental nature, and therefore needs to be rediscovered: the spiritual vision that allows us to see beyond the appearance of material things and, especially, beyond their usefulness or inconvenience to us, to perceive their God-given nature. This vision contrasts with the familiar, fragmented view that is precisely what makes the world an arena of human conflict: "we all see [the world] differently, in a way that is focused on separated selves."[17] Rediscovery of God's meaning in the world begins with work on ourselves: with a process of purification through the practice of virtues, an exercise of ascetic detachment that progressively liberates us from seeing everything solely in relation to our own preoccupations and desires. Ascetic writers often speak of a "practical" stage followed by a stage of natural contemplation; but it would be a mistake to think in terms of *first* attaining complete purification, *then* moving on to the next stage. The two processes fuel each other.[18] The more we are conformed to our own true nature, the more we are able to perceive the principles of nature in all the created entities we interact with (humans included), distinguishing the "*logoi* of things" from the fallen manner in which things exist. And the more we perceive creation in relation to its Creator, the more insight we can receive from it for the guidance of our lives, since all natures belong to the same pattern and are the work of the same hand. Learning from God's creation can start very simply: with wonder, with thankfulness, with lessons in living ("consider the lilies of the field"). But in this process, there is also a spiritual equivalent to "garbage in, garbage out." If one approaches created things without at least a desire for spiritual insight, one may well observe nothing more profound than the Creator's "remarkable fondness for beetles"—or, more dangerously, conclude that some form of social Darwinism is the law of life. This awareness of the ambivalence of natural phenomena in their actual mode of existence may help explain why the Orthodox notion of learning from nature never developed into a "natural law" tradition in the Western sense.

It should be emphasized that "natural contemplation" is not set in opposition to practical use of the world. "Alongside the utilitarian use of objects, *or rather by means of it*, one must learn to contemplate the flowering of heavenly realities in them," writes Olivier Clément[19] (italics are mine). Using the world in a contemplative spirit is one of the prime ways in which we learn spiritual lessons from it. The obvious example for the Church Fathers was working the land; but the same could be said of craftsmanship, or building, or scientific research. In order for us achieve insight into other natures and learn from them, they do not need to take the form of "nature" untouched by man: the nature of things can be obscured, but cannot be expunged. All that is needed is for us to recognize the things we touch as a reflection of the divine wisdom, not just raw material for our own uses. While there are plenty of counterexamples of humans using other natures in ways that are physically and spiritually destructive, there can be no opposition in principle between the nonhuman "natural world" and humans acting according to their own nature, which includes creativity.

The positive potential of human creativity is amply illustrated in Orthodox church life. As is often pointed out, the Eucharistic offerings do not consist of edible plants in their "natural" state but of bread and wine, sophisticated products of human culture, skill, and technology: it is in this form that the created world is offered to God in the Eucharist, accepted by him and given back as the gift of his own life. No, this does not mean that everything has to be turned into a human product in order to be worthy of God; it just shows us that bread-making on the part of humans is just as much a part of their God-given nature as is fermentation on the part of yeasts. And it reminds us that after all this complex production process, man still turns to God saying "All we have to offer is Your own gifts to us." Another example is the painting of an icon: wooden boards are sawn and sanded and prepared, minerals are ground into pigment and mixed with egg, in order to produce a depiction of Christ or a person in his image. The materials may seem to have lost their "natural" identity altogether; but looked at in another way, their natural identity is revealed. The tree, the rock, and the egg image the Word and Wisdom of God all the time—it just takes a laborious process before humans are able to see it.

Both of the above examples, the Eucharist and the icon, point us toward questions that are relevant when we start trying to assess new technologies. Does a technology or an invention enhance our sense that nothing can be claimed as our own, that all is God's gift? Does our use of technology highlight our awareness of dependence on God's gifts mediated through a host of other people and other creatures, animate and inanimate—or does it encourage a sense of human self-reliance? And in transforming the matter of this world into something that looks very different, does it make the world more transparent to the beauty, wisdom, and love of God, or less so?

What Bearing Does All This Have on Environmental Issues?

First of all, I believe that this cosmic vision of salvation liberates us from the crushing sense of facing a superhuman task with merely human powers. We are not suddenly being called on to "save the earth." Certainly, we face specific serious crises that compassion for other people and other creatures compel us to address urgently; but in spiritual terms, it is "business as usual." We are called to conform ourselves to Christ, and so participate in His work of saving all creation. This means ministering to the needs of others, if necessary at cost to our own comfort, in whatever circumstances we find ourselves (our actions therefore may variously be pigeon-holed as "environmental," "social," "humanitarian," etc.). It means ceaseless vigilance against the temptations of pride and pharisaism, of judging others and trusting in the righteousness of our own actions. It means faithfulness in even the smallest things, whether it is giving someone a cup of cold water or picking up a plastic bag that risks polluting the water. And perhaps most important, it means ruthless honesty about our failure to live up to this calling, and undiminished resolve, every time we fall short, to get up and try again. All this is what enables us to pray with confidence—to call on the only power that is able to make our own efforts efficacious. The wild discrepancy between the cosmic scale of the challenge (the salvation of all creation) and the mundane, apparently trivial character of the actions whereby I play my part in it, is a basic given of Christian life. The Christian has always been called upon to think cosmically and act personally.

So on the one hand, environmental crises are simply a part of the current context within which we are called to put love of neighbor into action. But on the other, Orthodox theological thinking unapologetically sees in such crises a spiritual message. This cannot be reduced to a moral message: it is not that environmental crises can be traced back to specific "environmental sins," though some have tried. Rather, like every crisis, danger, or suffering, it reminds us that we live in a world marked by a skewed relationship between humans and God. More specifically, it suggests to us that as long as we see God's creation exclusively or even primarily in relation to ourselves and our wants, we shall keep creating more problems for ourselves. And this is not because God is punishing us for transgressing a rule, but because we are going against our own true nature and that of all other created things.

To the secular mind, acknowledging environmental crises as a judgment upon ourselves is deeply unpopular because it is seen as leading only to guilt, to a sterile self-flagellation. The Christian approach differs in two important respects. First, the goal of such an acknowledgment is not guilt but repentance, a change of heart—the recognition that there is a way back, that we are not condemned constantly to repeat the same errors. The "nature" that we damage and degrade is first and foremost our own: it belongs to our true nature to act differently. But secondly, none of us can escape the need for repentance by shifting the responsibility elsewhere. As a matter of historical fact, some societies and sections of society have to date contributed to environmental destruction much more than others, but the underlying causes are rooted in the fallen human "nature" that we share, and its distorted priorities. This conclusion is only strengthened by the fact that more environmental ills result from structural factors than from culpable and willful damage by individual humans, and that so many problems can be traced back to seemingly "good" initiatives to improve our world.

The Orthodox Christian vision also makes a profound difference in our attitude to practical remedies. Particularly when we look at global warming in conjunction with the various other environmental threats, most of the realistic responses to the problems include reining in our own appetites—however many brave new technologies we may be able to wheel out. From the viewpoint of the Orthodox tradition, such a prospect does not mean being dragooned into a grudging restraint, deprived of "the good life" just when it seemed within our grasp. It means recalling ourselves to the ascetic way central to Christian life, the liberation struggle against our fallen "nature" and the tyranny of our own appetites. The ascetic tradition teaches what many of us know from bitter experience: that more possessions, more gadgets, more "consumer choices" only add to our dependence, multiplying the needs and cares that dominate our lives. Ascetic detachment is by no means to be equated with indifference or contempt for material things per se—quite the contrary. Restraint of our appetites breaks our usual dysfunctional relationship with the creatures round us, replacing it with what Olivier Clément calls "a wondering and respectful distance"—a respectful distance that extends even to the things that we have to "consume" in order to survive. No one should pretend that this process is easy, or that many will achieve it fully; it is perhaps one of the prime conflicts that we have to face. But there is nothing hypocritical in acknowledging and engaging in this battle with ourselves.

It is often remarked that societies, and individuals, usually start to worry about the less immediate and tangible environmental problems only when they have achieved a degree of affluence. This does not mean that environmental concerns are a frivolous luxury; it simply means that people living hand to mouth may have little choice but to accumulate debt, whether environmental or financial, and just hope that "something will turn up" in the future. But as Nordhaus and Shellenberger have perceptively pointed out,[20] willingness to face environmental problems may have more to do with people's sense of security than strictly with income: and it should be pointed out that there is more than one route to achieving security. It is notable that the Orthodox theology of creation is lived out most consistently in monasteries—not usually well-endowed ones—and above all in the lives of holy hermits who practice radical poverty. Such people have no financial safety-net at all: they live by the unshakeable conviction that the Creator who takes care of their friends and neighbors, the wild creatures, will take care of them as well.

"Preserving Nature?"

Where does all this leave "preserving nature?" When preparing this paper, I took "nature" to be a shorthand for nonhuman creatures/created things—though it seems with hindsight that it was meant to stand for a more abstract concept. In accordance with my original understanding, I suggest three principles involved in our treatment of the nonhuman natures with which we share the earth.

1. Created natures form the most universal "book" through which the Word of God speaks to human beings. Strictly speaking, the *nature* of created things, the reflection in them of the divine wisdom, is not something we are capable of destroying: we depend on the laws of nature for the workings of our most sophisticated technologies, just as much as we do in order to survive in the woods. But we are quite capable of distorting other creatures' mode of existence (or for that matter terminating it) so that they no longer *speak to us* of our common Creator. If we start with an instrumental view of other creatures and an arrogant attitude to the natural order, this has fateful consequences for everything we touch. Like toxins that become ever more concentrated as they move up the food chain, so our counternatural world-view becomes concentrated as we get further and further removed from obviously "given" natural processes. To change the metaphor, we increasingly find ourselves living in a hall of mirrors, where everything reflects our own fallen state. As our world becomes increasingly man-made and geared to our convenience, what we risk losing sight of is not a separate realm called "nature." It is the very nature of the created order, which means our own true nature as well—the fact that all created beings answer not to their own will, or ours, but to the will of the Creator.

2. Respect for the principle of other creatures' existence. The belief that all things are created according to a divine *logos* means affirming that their existence is not fortuitous or meaningless; it *makes sense* in the total scheme of things. Certainly, we are aware—as the Church Fathers were not—that extinction is a feature of the world as we know it: thousands of species have come and gone over the millennia. On the other hand, we cannot know when a species has lived its allotted span, any more than we can with a human individual. The knowledge that humans eventually die anyway is not considered justification for hastening their demise by deliberate action or

negligence; so should we not be similarly cautious about causing the demise of other species?

3. Our model for dealing with *all* creatures—and all conflicts of interest between them—is the "compassionate heart" of which St. Isaac the Syrian speaks: a heart that burns "for the sake of the entire creation, for humans, for birds, for animals, for demons, and for every created thing" and "cannot bear to hear or to see any injury or slight sorrow in creation."[21] This is more obviously relevant to individual creatures than to species; one cannot really exercise compassion toward categories. However, we also have as our example of God's relationship to his creation, as expressed, for example, in the Psalms: all sorts of creatures have their place, He has provided an appropriate habitat for each, and He "opens his hand, and all things are filled with good" (Ps. 104:28).

It is important to note, finally, that "every created thing" means precisely that. A compassion that is rooted in love for the Creator can never dismiss or devalue our human neighbor for the sake of nonhumans' creatures. St. Maximus points to the fundamental "natural law" that teaches us that humans are all "one nature"; that we violate our own nature if we do not care for the well-being of others as we do for ourselves.[22] But Maximus also speaks of a still more fundamental principle of unity: all creation is "one" inasmuch as it has a common origin in nonbeing.[23] "Creation" denotes a unity in frailty, but in a frailty encompassed and permeated by the Word who has created all things for salvation.

DEBATE

MATTHEW ENGELKE

Just a few observations on your presentation. One of the things that struck me has to do with the role of free will and human agency and the extent to which we should understand that concept vis-à-vis a divine plan. I think what you have also brought out is the very important issue of what we might call "the nature of nature," in the sense that you are here talking about a theological perspective in which nature means what an anthropologist would refer to as human nature—slightly different from the natural world. So these are just two general themes that we might want to consider. I have just one question, which has to do with the idea of divine presence in everything. I wonder how that should be understood. I say this because certainly within the Protestant missionary tradition that I have studied this is a huge problem for Protestant missionaries in nineteenth-century Africa, who were unwilling to accept local versions of an understanding of the divine presence because of its potential idolatrous nature. I wonder if you can just give me a sense, more sociologically than theologically, of how these kinds of issues play out within Orthodox societies.

ELIZABETH THEOKRITOFF

As to the question of divine presence: one very interesting insight on this comes from the Orthodox mission to Alaska in the eighteenth-nineteenth centuries, when Alaska belonged to Russia. The missionaries working in Alaska were generally very sensitive to the local culture; and they were readily able to see in the various divinities or spirits that the native people worshipped that these were precisely *logoi*, "words" of the divine Word. These were reflections of the Creator's word, and the process of evangelization was therefore rather like St. Paul talking about the "unknown God" whom the Athenians were already worshipping unwittingly. There are also parallels in more recent missionary work. The idea of "words of the Word" in creation does make it much easier to be open to manifestations of God in traditional religions and particularly to "baptize" material expressions of religious faith—something that the Church did from a very early date—without a morbid terror of idolatry.

"Has there been reason to argue over what constitutes divine presence in everything?"—Not as such, to my knowledge; or not for many centuries. There was the Palamite controversy (fourteenth century) over the vision of divine light, which had to do with how we maintain both that God is unknowable, and that he is accessible to material creatures; this was resolved in favor of Palamas' doctrine of divine energies in all creation. The energies are God in action, yet not the divine essence. Again, one could say that the iconoclast controversy (eighth to ninth centuries) had to do with the way in which God is imaged in his creation and can be venerated through material things.

One clarification about "nature": I think you suggested that I identified nature primarily with human nature, which is not what I was saying. I did say that in Patristic usages "the nature" without further qualification is used in a concrete sense to mean the human race. But my main point is that in the general (more abstract) usage, "nature" refers to the nature of everything that is, including humans; the way things are created to be.

BRUNO LATOUR

I have many questions. One is the clarification of the concept of "nature" that you use. If I translate it with "vocation" would that work? It is not "nature" in the sense that we have been interested in here, and that relates to the destiny of science or to the continuity of organisms. When you say that nature cannot be destroyed, but our human nature can be destroyed, if I translate nature with vocation will that work?

ELIZABETH THEOKRITOFF

No, I was not saying that our nature can be destroyed. What I wanted to say was that if we get too far from "nature," in the sense of nonhuman creatures, entities that have not been shaped by humans and the way these creatures function, then we risk loosing sight of the fact that we live in a world that is the work of a Creator and not primarily the work of ourselves. In that way, we are also loosing sight of our own nature. We are not destroying it, but we are forgetting that we are creatures—a reality essential to our own nature.

BRUNO LATOUR

So, is it "nature" as our destiny? It is not "nature" in the sense we have discussed it previously. There is no cosmology implied in the notion but the vocation of developing one's own destiny.

ELIZABETH THEOKRITOFF

I have to say that I never quite understood in what sense some participants used "nature." I would say, however, that "the nature of created things" in a theological sense is by no means unconnected to science. The world through which we perceive God's "invisible power" is the same world that the sciences study; and what the sciences reveal about things are aspects of the nature of those things, even if not the totality of their nature. As to "nature" and "vocation," it seems to me that these are rather different things. Certainly, if we return to the passage that I quoted from Bishop Kallistos Ware, where he describes the *logos* of a things as "God's intention for it, the inner essence of that thing, which makes it to be distinctively itself and at the same time draws it towards God," then we see that in a sense the nature of a creature—what it is in itself—encompasses its vocation. And this applies to all created things, not just humans. But "nature" is not a synonym for "destiny." By nature man is a creature—a created being, in contrast to God—yet one created "in the image" of God. This aspect of his nature, the in-the-image, defines his destiny as "an animal called to become God," in the phrase of St. Basil the Great. He cannot attain this if he forgets where he is going (his "vocation"), nor if he forgets the creaturely aspect of his nature.

BRUNO LATOUR

I was very struck by this beautiful argument that we are no longer alone in the task of environmental transformations because God is helping us. James Watt (he was the Minister of the Interior to Reagan, if I remember) said that there is no need to save the forests or the mines of America because the Savior is going to come in a few years anyway and thus there is no reason to protect all those earthly things. So how would you, in this discussion here, accommodate the third term that is God here as a helper in those disputes? Because God has quite a damning track record of contributing to destructions as well. So, it is a strange God that could help us protect the environment and he is also the one who helped Watt to say that "the environmental movement is of no importance, just let's exploit the planet because in the end it has so little importance."

ELIZABETH THEOKRITOFF

Well, I am not sure what evidence we have that God did "help" Watt; the planet has not yet been destroyed, after all. As I said, there is a parallel between created nature as a whole, and Scripture: and one aspect of the parallel is that both can readily be misused, and both can be misread. Certainly we see this with Scripture; people do many strange things ostensibly in the name of God. That is why I also emphasize that the preparation for rightly reading both Scripture and nature involves working on ourselves; essentially, it is an ascetic preparation that brings us closer to Christ. St. Paul talks about having the mind of Christ; and we know that when a book or a letter or a speech is hard to interpret, our best hope is to "get inside the mind" of the author. This often enables us to say with a fair degree of confidence that he could not have meant this or that; it is not in character. One of the frustrating things in Christian life, however, is that we never have the definitive answer until the end. Throughout our lifetime, we never know for certain whether we are on the right track, whether we actually are furthering God's purposes or working against them. We can make every effort to be on the right track; but this does not mean jumping to conclusions about God's plan and then developing our own ideas for how to get there. It involves trying to discern God's will at every step of the way. We can strive to listen, and we can try to teach ourselves to listen; but we can never be certain.

BRUNO LATOUR

Beautiful answer to the difficulty of interpreting the book of Scripture and the book of Nature.

ANDREA VICINI

In reflecting on environmental issues, I ask myself: How can we respond theologically to current environmental problems? How can we understand God's way of acting in today's world? How does God act in the world? How would you characterize the specific contribution of the Orthodox church when we try to understand the centrality of the relationship between God and humanity? Do you notice any different insights when we consider the past Orthodox tradition or when we focus on recent theological developments within the Orthodox church?

When we look at the Catholic, Protestant, and—it seems to me— Orthodox theological reflection, we recognize different ways of articulating and of expressing the interaction between God's action and human action. Do you agree? Do you recognize any Orthodox specificity? It seems to me that there are different emphases and nuances. How do we think about God, humankind, and human action among Orthodox, Catholics, and Protestants?

ELIZABETH THEOKRITOFF

One thing that I think is really important is that one and the same God is seen as acting in the workings of the natural world and in history: there is no sharp distinction between these different aspects of God's action. On the other hand, it is exceedingly difficult to articulate God's action in creation. One cannot speak of it as "intervention"; there is no "outside" from which to "intervene" because it is totally God's creation. In rather the same way, you cannot talk about a novelist "intervening" in the plot of a novel he or she has written; or if you can see the author "intervening," if something has evidently been tweaked just to make the story work, then it is a badly written novel. That is inexpert writing, and that is not the way God works in creation. The understanding of God, of the Word, present in the principles (*logoi*) of creation is, I think, the basis for trying to speak about how God operates in creation and how, in a certain sense, he is the working of creation. He is the principle of its nature, which, as I have said, is by no means always identical with its mode of functioning, its mode of existence. Hence there are things happening in the nonhuman world, as there are in the human world, which one cannot see as a direct reflection of God's will.

It does seem to me that there are some quite marked differences of emphasis among (and within) the Christian traditions. I am always uneasy about making comparisons because I do not feel that I know other traditions well enough to avoid caricaturing them. But one key notion in Orthodox thinking is that of synergy between God and man; we are "coworkers with God." God works in and through us, just as he also works in and through the natural world. There is thus no contradiction between saying "only God can save us" (from environmental disaster), and offering ourselves as his instruments, doing everything possible to put our energies, our resources, our skills, and our technologies at his disposal for the purpose. But again, the imagery I am using risks being misinterpreted in an external sense, as if God were the overseer of a construction project. Human action in the world is to be not just for God, but in God, in relation to him; so it always has a "triangular" quality. We offer our work to him, and he does with it as he wills.

A striking example of synergy is God's central action in the world—his incarnation, which he made dependent on the Virgin's assent. And it is important, I think, that the Orthodox tradition affirms strongly that the Incarnation is not only for the sake of man, and not only to redeem from sin. It is to bring the entire creation to fulfillment. But that requires man to stop seeing the world only in relation to himself, and instead relate it to God. The sacraments provide a sort of template for this; they offer things of the earth, whether shaped by humans or not, so that God can use them to convey his blessings. But I am not sure if this is an adequate answer to your question.

IZABELA JURASZ

My first remark concerns the exchange between Bruno and Elizabeth on the subject of the possible destruction of human nature. I think we might speak more of the destruction of the passions of nature and not so much about the destruction of human nature itself. The ancient philosophical tradition and also Patristic theology mention the destruction of those passions in nature that bring human beings closer to animals. Passions are not only morally evil, however. They are most importantly a sign of change. To be driven by passion means a change is taking place. Human beings must thus either combat evil passions or transform them. The aim is to perfect human nature and definitely not to destroy it. This was just a little comment on the previous exchange.

Now I'd like to ask Elizabeth a question. On listening to your discussion with Andrea, I wondered what you thought about the notion of the Fall: how does it fit in with the theology of nature? The differences between Orthodox Christians, Catholics, and Protestants would seem to be in relation to this issue. Can you explain a little further how you understand this notion of the Fall in your thinking on human nature? I also wonder if you need this notion or how it can be interpreted differently in Augustinian-inspired theology, which is more Western than Eastern. In fact—and I'm simplifying in the extreme—in Catholic and Protestant theology the Fall is much more important than in Orthodox theology. For some—such as Augustine, Anselm of Canterbury, or Luther—the Fall is a major event that marked human nature. But for others, and I'm thinking of the Greek Patristic tradition, the Fall is only an incident and is not of crucial importance. In terms of your talk, what do you think about the Fall?

ELIZABETH THEOKRITOFF

Yes, I take your point that the traditional Western view certainly presents a much gloomier picture of fallen human nature than the Eastern tradition does (the Eastern emphasis on synergy between God and man is indicative of this). But this does not make the fall a minor incident. St. Paul tells us that through sin, death came into the world, which is hardly a trivial matter.

The point is rather, I suggest, that the consequences of the fall are seen in the East in existential rather than moral terms. There is certainly not the same stress on "original sin"; but when we start talking about the cosmic effects of the Fall, I cannot by any means agree that this is unimportant in the Greek Patristic tradition. There is a very strong tradition that the way we live is marked by a Fall. I am not talking about our bodily existence—that is innate to us, part of how we were created to be—but the way we relate to the world, and particularly what in the ascetic tradition are called the passions, our impassioned relationship to things and people. This relationship is interesting because one of its key features is a dysfunctional love of the world; we "love" the things of the world in the sense of wanting to possess them, sometimes with an insatiable appetite, or wanting to get them for ourselves before somebody else does. This whole attitude to the world is seen as indicative that something has gone wrong. And furthermore, the Fathers believed that what is known as natural evil is somehow connected with the fall; humans were intended to have power over the elements, and indeed we sometimes see evidence of such power in people of great holiness. Whether the connection between fall and death means that mortality in nonhuman creatures is a result of the fall of man is very much a moot point; for a number of the Fathers, clearly it is not. There would have been death among nonhuman creatures without any fall. But that does not address the question of whether there is an eschatological condition in which in fact death and corruption are abolished for all creatures; that is a possibility that is still open. So to try to answer your question: Does our theology of creation require the notion of a fall? As a historical event, certainly not—it only confuses the issue. But there is a very solid tradition in the Christian East as much as West that (1) the present state of the "natural world," including death, does not correspond to God's ultimate intention for it; (2) attainment of that ultimate state depends on man's relationship to God, and something has gone wrong with that relationship. That relationship is restored in Christ, but the restoration is fully manifested only on the other side of his death, in the resurrection. I believe that this tradition is very important to our understanding of how God's goodness operates in a world that often strikes us as anything but "very good" (e.g., we keep coming up against real conflicts). And conversely, it helps us to understand how the state imaged in Eden is a reality promised to us, not just a nice "golden age" myth, despite the obvious gulf that separates it from our everyday experience of the world. To come back to your first question: this was about the passions of our nature?

IZABELA JURASZ

Yes, indeed. I think for instance at Gregory of Nyssa and at his treaty on the creation of man. He seems to me very representative of the Patristic tradition. In this work, he speaks about the relationship between man and animals and discusses also the destruction of passions, which actually is not a destruction of human nature.

ELIZABETH THEOKRITOFF

Absolutely. Nobody advocates the destruction of our nature; if our nature is distorted or obscured, that is unequivocally a bad thing. As you know, the Fathers are not unanimous in the way they speak of the passions; some do see the passions as being extraneous to our true nature and therefore needing to be destroyed, while others see them as part of our nature and needing to be transformed, restored to their proper function. But at least in the way we experience the passions, they are related to our mode of being, which is a fallen one, rather than the principle of our nature, which is God-given. In fact, that is one reason that from the beginning I have felt a bit uncomfortable with the title of the Dialogue because I do not know what these "religious passions" are; I am more used to the term "passions" being used in a negative sense.

BRUNO LATOUR

I'm interested in seeing some terms used simultaneously in theology and in ecology: Sin, Fall, and Harmony. Harmony corresponds to the pre-Adamic state, and it is retranslated by the ecological movement as the equilibrium of pristine nature. And Sin is reinterpreted as humans "spoiling their nest." What sort of new energy would be given to the environmental movement if Fall and Sin and Harmony were recycled explicitly from the old tradition and would no longer be limited by their use in environmental ethics (which have no base in the real world, where harmony is just a misinterpretation of the past)?

ELIZABETH THEOKRITOFF

The relation of the Fall and Sin and Harmony to environmental questions is a very interesting point. It seems to me that in our fallen state, our human aspirations to be divine and in this way achieve our proper greatness are not focused toward God but have their focus within this world. So we try to create the nearest thing to an earthly paradise, which for some people means amassing wealth and possessions and making life as easy as possible. And the "back to nature" argument, representing one particular variety of environmentalism, is the other side of the same coin. It is a very different sort of earthly paradise; but again, it is talking entirely in terms of this world. So instead of seeing the perfect harmony of creation as something essentially eschatological, something that we have to work toward by conforming ourselves to Christ, harmony becomes something that we can find in this world if only we get the right system, if only we make the appropriate social changes. And correspondingly, the Edenic, pristine relationship with nature is placed in this world; it is seen as an alternative organization of society. On the other hand, most Orthodox do believe that the environmental crisis carries a spiritual message; but it is not a message about specific environmental sins. Like most crises, this one reminds that there is something wrong with our relationship with God. I also mention in my paper that the idea of the Fall is not very popular: not only because people consider it too mythological, but also because, if you believe that something radical has happened that differentiates this world from the Garden of Eden, this means that no amount of social change or environmental programs or indeed technological progress is going to produce a sort of approximation to Eden. We do not live in a world that can be turned into paradise by our own efforts; and if we recognize this, we will make less exacting demands on the world. As

St. Ephrem the Syrian says, "In the world there is struggle." We can improve things, but we have to recognize that there really are limitations.

As for sin, I find this something quite problematic to try to introduce outside an ecclesial setting because it is so widely and profoundly misunderstood. It is associated in the secular mind with transgression of moral rules or violating a puritanical code of conduct (so something enjoyable can be described as "sinfully good"), and with the worst sort of caricature of religion, an image of a god whose main occupation is punishing people for infringing rules. (All this is discussed at length in a very interesting book that I translated some years ago: *The Freedom of Morality* by the Greek theologian Christos Yannaras.) And, for reasons that I find hard to articulate, the idea of "ecological sin" makes very little sense to me. Perhaps it is because the chain of personal cause and environmental effect is usually so convoluted; ideas of individual moral responsibility really don't seem to fit very well with the sense in which humans are responsible for large-scale environmental problems. I do believe that environmental crises have to do with human sin; with our fundamental sin of turning away from God, making earthly goods ends in themselves. This means that nothing in the world works as intended; that not just "bad" actions, but all our striving for good things, starting with life itself, just creates more problems for us. But this understanding of sin makes sense, and is bearable, only in light of the Savior.

SIMON SCHAFFER

This is a gloss to your notion of garments of skin, the rather troubling account of the second nature of human technique in society. Is there a set of ways of telling the difference between techniques that belong to second nature in this fallen world? You gave us a list: art, skills, technology, learning, science, law, culture and politics, and a set of techniques that really aren't like that, baking the bread, making the wine, producing the icons, which are part of the Eucharistic project. Is it conceivable that there might be technologies not currently much practiced that could switch usefully from the order of the second nature to the order of the Eucharist? There may be technologies that we do not currently much practice that would no longer be signs that we live in a fallen world because all they provided us with is ways of coping with inevitable imperfections.

ELIZABETH THEOKRITOFF

I think the trouble is that I did not explain this fully enough. I am now starting to understand better where Izabela's question was coming from. I was trying to talk in a very matter-of-fact way about "the world of the Fall" being qualitatively different from paradise; but people seem to be hearing it in a doom-laden (perhaps Augustinian?) sense, as if everything were tainted by sin. But this world of the "second nature" is not tainted by sin;

it is a world precisely adapted to our state (which *is* marked by the sin of turning away from God and making the world an end in itself), arranged in such a way that our mortal state with all that it entails becomes our path to salvation. So, as to the question of whether there could be techniques that do not belong to the "second nature" in this fallen world, the answer has to be No. This side of death, the world as we know it is all we have; that is what we live in, that is the world that Christ came into. And that is the world in which we celebrate the Eucharist: the "Eucharistic project," as you happily term it, is not about a different set of techniques but about transforming the techniques that we use every day, "making them other" in the words of the poet David Jones (*Anathemata*).[24] The technologies that I mentioned—or that Nellas mentions—were not intended to form an exclusive list. It was some examples of things that can be used for good; these include things that can make life better in material terms, as well as the technologies required to produce the Eucharistic bread. Given the world we live in, those technologies and natural processes are the way in which we make Christ present in this world. The point here is that all these things are given so that they can be used to sanctify the world, to realize and manifest God's presence in it.

SIMON SCHAFFER

I wanted to get clear whether those two lists are finite in number. Could one envisage a development in the history of grace where absolutely novel technologies appear, or is the list closed now?

ELIZABETH THEOKRITOFF

In order to answer that question, one has to try to imagine the eschatological state. I am working on the assumption, for which there is some basis in the Church tradition, that the Eden story can be seen as to a great extent eschatological. It is a picture of where we are meant to be going. One may point out that Adam was told to "till and keep" the garden; but it does not seem that technology loomed very large in paradise. Similarly, in the resurrection I have no idea what role, if any, technology will play . . . this sort of speculation quickly gets rather silly. I think the main point is that as long as we are on this Earth, there is only one list, not two. There are most definitely ways in which all of this—culture, technology, and so forth—can be used in such a way as to manifest God's presence rather than to obscure it. For all I know, absolutely novel technologies could certainly appear; but they will still be techniques of this earth, not of the new creation. And they will challenge us with much the same fundamental questions as the plough and the oven did in their time: do we see these advances as serving an autonomous end, merely giving us a more comfortable life in this world? Or do we see them as ways of offering the world to God, serving him in other people who are in need, glorifying him in his creation?

Notes

1. See Chapter 9.
2. See further Elizabeth Theokritoff and George Theokritoff, "Liturgy, cosmic worship and Christian cosmology," *The Messenger* (*Journal of the Episcopal Vicariate of Great Britain and Ireland*) 11 (August 2009): 15–34; corrected version in *Sobornost* 32:1.
3. Bruce V. Foltz, "Nature's other side: the demise of nature and the phenomenology of givenness," in *Rethinking Nature: Essays in Environmental Philosophy*, edited by Bruce V. Foltz and Robert Frodeman (Bloomington and Indianapolis: Indiana University Press, 2004), 334.
4. *Ambigua,*P G9 1:1077C.
5. Bishop Kallistos of Diokleia, *Through the Creation to the Creator* (Pallis Memorial Lecture 1995; publ. *Friends of the Centre*, London 1996), 11.
6. To Thalassius 2; Paul M. Blowers and Robert Louis Wilken, trans., *On the Cosmic Mystery of Jesus Christ: Selected Writings from St Maximus the Confessor* (Crestwood: St Vladimir's Seminary Press, 2003), 99–101.
7. This is Thunberg's characterization of the teaching of Maximus; see L. Thunberg, *Man and the Cosmos: The Vision of St Maximus the Confessor* (Crestwood: St Vladimir's Seminary Press, 1985), 75.
8. Fr. Dumitru Staniloae, *The Experience of God: Orthodox Dogmatic Theology*, vol. 1, *Revelation and Knowledge of the Triune God* (Brookline: Holy Cross Orthodox Press, reprinted 1998), 1.
9. Ibid.,2 1.
10. Foltz, "Nature's other side," 336.
11. To Thalassius, PG 90: 481C.
12. Cf. Maximus, *Ambigua,*P G9 1:1349A.
13. To Thalassius 59; PG 90: 613D.
14. Christopher C. Knight, *The God of Nature* (Minneapolis: Fortress Press, 2007), 87,1 36.
15. See Panayiotis Nellas, *Deification in Christ* (Crestwood: St Vladimir's Seminary Press, 1987), 43–91.
16. Cf. Gregory of Nazianzus Or. 14.20, quoted in Nellas, *Deification in Christ*, 91.
17. Andrew Louth, *Maximus the Confessor* (London: Routledge, 1996), 37.
18. Cf. Kallistos Ware, *The Orthodox Way* (Yonkers, NY: St Vladimir's Seminary Press),1 06.
19. Olivier Clément, *The Roots of Christian Mysticism* (London: New City, 1993), 220.
20. *Break Through: From the Death of Environmentalism to the Politics of Possibility* (Boston/New York: Houghton Mifflin, 2007), for example, 149–150.
21. Homily 71, *The Ascetical Homilies of St Isaac the Syrian*, translated by Holy Transfiguration Monastery (Boston: Holy Transfiguration Monastery, 1984), 344–345
22. To Thalassius 64; Blowers and Wilken, *On the Cosmic Mystery of Jesus Christ*, 167–169.
23. Ambiguum 41, PG 91: 1312AB; Louth, *Maximus the Confessor*,1 60.
24. D. Jones, *Anathemata* (London: Faber & Faber, 1972).

7

Seeing the World, Hearing the Word

Cosmos and Creation in Protestant Theology

Anne Marie Reijnen

What I really want to concentrate our attention upon, and to contemplate, is the vision of the "blue orange" (Figure 7.1), which I suggest is a powerful image that has now become a common place and a commodity. Yet, it retains the power to attract and to intimidate. It fulfils then the dual nature of the Holy as defined by Rudolf Otto, the *fascinosum* and *tremendum*. When I speak as a Protestant theologian of the theology of creation in the service of ecology, what I find most compelling is this image of the "blue orange," the image of the Earth that was beamed back to Earth by human beings. Now this actual image beamed back to us from space was the validation of early representations of the Earth as a sphere, such as Atlas, the image of God as Pantokrator, *omnitenens*, and the visions of Julian of Norwich. When I speak of cosmos, my purpose is to elaborate what is a more constructive term for this time and place; I think it is cosmos (rather than creation, for example): this can serve to define a common ground between Christians and non-Christians. This choice does go against the grain of a large part of my own tradition, as I will now try to show and explain. *Kosmos* implies an appeal to the sense of seeing; *kosmos*, as order and beauty, are at loggerheads with the great emphasis placed by Martin Luther on the sense of hearing, the logo-centric emphasis of the sixteenth-century Reformation. Martin Luther said that we are here to learn from the living word of God, more hearers than seers of God's world, or cosmos.

We could call the project of the Reformation "a theology of hearing the Word of God." It tends to make reflection about nature suspect. Speaking and hearing is how humans interact with each other and with God, whereas humans interact with nature by seeing or by contemplation. So what I am trying to show is the conflict, within the Protestant tradition, between seeing and hearing and the priority of hearing over seeing. "Moreover the gift of the Word is the most valuable" said Martin Luther, "if you take this away

Figure 7.1 TheB lueM arble.
Source: http://en.wikipedia.org/wiki/File:The_Earth_seen_from_Apollo_17.jpg

it is like taking the Sun away from the Earth." This is really a theology that is centered on the receptive, passive hearing of the Word of God, but the Word is of course itself a dynamic reality; it is very much like the Hebrew *dabar* that is both a word and an action, it is dynamic. What do you find when you see? Typically, in the Lutheran paradox, what you get or what you see is only the *deus nudus*. The God of hail and storm and thunder is for all to see, the *deus nudus*. For Luther, this is a different experience from encountering God who is revealed through Scripture and in Christ. God is hidden in God's manifest majesty. The *deus nudus* is in fact the God who "hides" godself (*deus absconditus*). In fact, you do not want to see that God, for you might well be destroyed; if you hear the Word, on the other hand, you will be healed. Now, despite this attack on seeing, Luther is ambivalent and therefore interesting, for he does have a "rich apperception of God throughout the world of nature."[1] Martin Luther invites us to marvel at a grain of wheat. If we really understood it, he writes, that one single grain of wheat, we would die of wonder! There is an internal complexity there, an ambivalence that can of course be exploited by theologians. Luther affirms strongly the materiality of the sacramental elements, the bread and the wine of the Lord's Supper and the flowing water of Baptism. So while Luther is pessimistic about human nature, at the same time he affirms that the finite human body (by definition mortal and finite) is capable of the infinite, *finitum capax infiniti*.

One generation later, John Calvin is stressing, by contrast with Martin Luther, the importance of seeing. Calvin would say: the invisible God makes God himself visible in many ways: the created world is the theatre of the glory of God, *teatrum gloria dei*. Also in the Scriptures and in the face of Christ; those are the three ways in which the invisible God becomes visible. Calvin's favorite image of nature is a sort of mirror in which we contemplate God who otherwise is invisible, but it is true that for Calvin to truly appreciate this mirror or this theatre—there are two metaphors there—you do need the spectacles or glasses of Scripture. So the emphasis on seeing still remains very logo-centric.[2] Now both Luther and Calvin held that nature itself has been cursed by God in response to sin. The Earth is now less fertile than it had been and some animals have begun to prey on other animals. What is problematic for us in the twenty-first century with the heritage of the Protestant Reformation and the Magisterial Reformation of the sixteenth century is the intensely anthropocentric understanding. That is really the problem, the anthropocentric understanding, and here is the clear conflict with the "biocentric" approach of today's Protestant eco-theology. One quote: "The end for which all things were created was that none of the conveniences and the necessities of life might be wanting to man," this is Calvin. So you can see the created world as the storeroom and Luther even uses the German word *Inn*, the Inn, the hotel if you will, for human beings, conveniently stored with all commodities of human life. Calvin stressed the importance of the *vita activa*, active life, over the contemplative life, *vita contemplativa*. The believer is called to transform the world, to work for new social forms and novel configurations of the human environment. For instance, Calvin believed that you do not see the face of Christ in the poor, rather you have to work to eradicate poverty. So there is a confidence, born of the sense of the divinely ordained historical destiny. This in turn stimulates the later development of Puritanism, and the idea of the American exceptionalism. They reflect Calvin's belief that human beings are called, in the wake of the creative activity of God, to transform the World.

As we talked about literary figures that are important for us, we mentioned Frankenstein; I would surely add Faust. But I would say that the hero, the protagonist of what I just described as the "Calvinist belief in human capacity and vocation to transform the World," is Robinson Crusoe. Robinson transforms the inhospitable island with a tool in one hand and the Bible in the other hand. The spirit of Calvinism sets the stage for secularized developments in the Western industrial and mercantile society. So it is not fair to say that this is the spirit of capitalism, but it surely is conducive to the spirit of capitalism. I would like to say, continuing on Calvin, that the influence of Calvin on the English-speaking world has been much far more important than that of Martin Luther. The scientific worldview formulated in the seventeenth and eighteenth centuries assigned an important religious role to nature. Nature provides the grounds for believing in the existence of God. Calvin had also said this, but he had restricted it to what believers could learn about God from Nature. Now in the seventeenth and eighteenth centuries a natural theology is developed that pretends or hopes to do the same, but argues only on the grounds of reason. Reasonable Christianity believes that in nature, through reason, we can discover the Maker. In the work, in the artifact, in the complexity of that artifact, human artifact compellingly calls for a Maker and that is really "natural theology."

Descartes is of prime importance, the Cartesian distinction between mind and matter. The partition between thinking substances and unthinking *res extensa* becomes an ally in the natural theology that is of Protestant origin. So I am not trying to explain Deism in the English-speaking world; I will only point out that it is not the same as French-speaking *déisme*. The laws of motion are imposed by the supreme law-giver on Nature; Nature is understood as entirely passive. And there is an analogy between the physical world and the world of morality. The same Law-giver who has laid down the laws on the physical world, the world of matter, has also laid down the laws for human behavior. This second source of Protestant religious understanding leads no more to ecological sensitivity than Calvinism does because Nature is now viewed as a machine made by God.[3]

Now fast forward, a brief comment on pietism. Pietism puts a premium on religious experience; it emerged as a reaction to the rationalism of much of the Protestant tradition. The Romantic movement and *Sturm und Drang* is yet another reaction. Finally, after the Second World War one witnesses the ecumenical development of "eco-theology." I would describe this very briefly, again using the overview provided by Paul Santmire and John Cobb in the book edited by Roger Gottlieb. The World Council of Churches was confronted in 1961 by a Lutheran theologian called Joseph Sittler with an image of the cosmic Christ. This was immediately decried and put down because in 1961, for Lutherans, one was still too close to the Nazi ideology of blood and soil and of vitalism. So this proposal by Joseph Sittler of a return to the ancient vision of the cosmic Christ could not be heard in 1961 in New Delhi. The Assembly held in Nairobi (1975) was a turning point: at this point, the World Council of Churches, most of whose members are Protestant churches, developed the theme of sustainability. Earlier, there had been the quest for a just and participatory society; in Nairobi, a third element is added, which is "sustainable society." In 1975, the delegates were still fully within an anthropocentric logic: sustainable is understood to be "sustainable in order for the human community to live in." That is why the Earth had to be sustainable, that is why society has to be sustainable and not only just and participatory. Another shift occurs in Vancouver in 1982. At that time the delegates from the First and Second worlds—Europe and North America—are in the midst of fears for a Cold War. So their agenda is peace. But they have to accept from the Third world the quest for justice. So we have peace and justice and then someone, I do not know who, but it is accepted immediately, suggests to add "integrity of creation" and this truly is a turning point. In Vancouver (1982), the three goals are concomitant: justice, peace, and integrity of creation. From here on, we really witness a shift within (mostly) Protestant theology in the English-speaking world. Over several decades, a considerable body of works in the new field of "eco-theology" has been constituted. They carry revealing titles: *The Body of God* and *A New Climate for Theology* by Sally McFague; *Eco-Theology* by Celia Deane-Drummond; *A Greener Faith: Religious Environmentalism and our Planet's Future* (edited by Roger S. Gottlieb); *Earth Community, Earth Ethics* by Larry L. Rasmussen.

Ecology is seen as a new field for human vocation, this time with a modified anthropocentrism. One perceives this new climate in, for instance, in the Lutheran theologian Paul Tillich. In 1963 already, reflecting on "Life and Its Ambiguities," he wrote: "The inviolability of living beings is expressed in the protection given to them in many religions, in their importance for polytheistic mythology, and in the actual participation of man in the life of plants and animals, practically and poetically."[4] We are gradually moving toward a biocentric approach in theology.

DEBATE

MATTHEW ENGELKE

What I particularly appreciate about this presentation is the introduction of the history of the senses into this all. It was very interesting for me to hear. You emphasize the Protestant emphasis on hearing especially given the extent to which that was squashed in so many ways, certainly in nineteenth-century America. I do not know if you are familiar with Leigh Eric Schmidt's book, *Hearing Things*, and ways in which Protestant intellectuals and actors in the United States tried to suppress the senses of sound with the shift to visuality. This is a side point, but of course certainly relevant to our concerns here, the history of the senses.

MICHAEL SHELLENBERGER

This is great. I feel this experience that when you hear a paper and you realize that what you have written before is something that you wish you've written differently. In the traditional, conventional environmental discourse, viewing the Earth from outer space was a watershed moment: this was when we realize that the Earth was fragile, that it was small, that we should not be hubristic, it was actually a call against hubris. But this picture is indeed the result of an incredible act of hubris: to put a rocket space into motion, to develop the camera lenses that could photograph the Earth, it is active, you are penetrating the atmosphere. Instead, we have turned the Earth into a fetish, it has become a substitute: we are not seeing the Earth but we are seeing the photograph of the Earth (one depended on huge amount of technological development), and this becomes a call to passivity while everything that allowed that call for passivity was the result of an active and aggressive process. I would love to hear your thoughts about what you see as giving rise to this particular moment because I think you'd reject a kind of naive view: we saw the Earth, we joined hands, we sang "Let it be," we read Heidegger and we decided to save the planet.

ANNE MARIE REIJNEN

Oh no, it is much more complex than that; that is why I used the *mysterium tremendum* of Rudolf Otto. I appreciate the point that you make, I would have added one more political point that is cosmopolitism. To view the Earth

Figure 7.2   Woman holding globe on her hands.

from outside is a stark reminder that this is the one place where we live with other animals, with the plants and the rocks and the ocean, so that on the positive side. On the other hand, I refer to the Atlas complex: remember Atlas carries the world on his shoulders; this is a men's world, it is not only anthropocentric it is also androcentric, so this is really a man carrying the world on his shoulders and he is a white man if you combine that with Rudyard Kipling's *The White Man's Burden.*

We have a trajectory that goes from the image of God confessed as the *omnitenens*, the omnipotence of God (and this was wrongly translated into the God Pantokrator, who holds the globe in God's hands) to the notion of the human community working for the sustainable Earth. You see the burden has shifted: it is human beings who are now "holding the Earth." You see here, in this image (Figure 7.2) a human hand holding the globe. It is a powerful image and it is completely ambivalent.

MICHAEL SHELLENBERGER

Well, I know that prior to this act of caring for this orb is an incredible amount of human power. If you really want to let being be, then you would not be putting your body in front of tractors erasing the Amazon, you would not be accelerating technology and innovation to have clean energy, if you really want to let being be then you would really retreat with Ramana Maharshi or to the Black Forest.

TED NORDHAUS

This is a godly perspective, now we are viewing it in God's perspective. In our book we quoted Stewart Brand, who became famous for putting that image on the cover of his catalog, *The Whole Earth Catalogue*, meaning "We are as Gods, we might as well get used to it." This might be the real meaning of this image, as opposed to the image of a small, fragile planet.

BRUNO LATOUR

My question has to do with the idea that we have opted for *kosmos* as the common ground between Christians and non-Christians, and that there is a risk involved but the peril of acosmism seems to us to be more harmful than the risk of "panentheism." This "we" is actually Anne Marie, it's not the World Council of Churches. So, firstly, I would like the definition of "panentheism" and also I would like to understand if this is the end of the trajectory that you briefly summarized because it means a redefinition of the transcendent. And the codicil of my question is: does it allow Calvinism and the Protestant tradition to reconsider their constant accusation against the pagans of being too much connected with the *kosmos*?

ANNE MARIE REIJNEN

There is a sense there on my part of theological injury that has to be healed by the same theology. I mean there are two ways to go about this: you can say "This is where theology of this brand has brought us and so I turn my back on it" or you can say "Maybe the remedy is in the disease" so then you go back to the body, *le grand corps malade*, and you rediscover the smaller voices, the voices on the margin but within that tradition. For instance, in Luther I find, apart from his pessimism about human nature and his unreasonable diatribe against reason, much to feed a theology of nature and I briefly alluded to that, the sense of wonder, the trust in the goodness of nature, of the world, of things, of being and having a body, all of that is in Martin Luther. So the main trust of the Reformation was indeed against idolatry that was conveniently identified at the time with the Antichrist, who was at that time in Rome…Within our tradition there is a ground for building a view of creation that is a "modified panentheism," not a pantheism, that is not the belief that a tree or a bush or a river is God or godlike, but that God is present in the world (Panentheism). So that is my heresy, heresy meaning choice: that is my choice, that is my heresy.

ELIZABETH THEOKRITOFF

I want to come back to Michael's point about the ambiguity of the image of earth from space because it seems to me that this really throws light on something that we were discussing a short while ago: the way that we can risk loosing touch with nature, in other words the nature of all beings, through technology. This is a risk that we run, but it is not something that is inevitable. Here you have, as you both pointed out, a striking image of the Earth that depends on an enormous amount of technology. And there are two totally different ways of viewing this: we can say that because of the technology, which is also God-given, we are able to see this image, and what we see is something small and fragile and dependent on God and something that we are all in together, all humans and all other creatures of the Earth. Or you can look at it and say that what this tells us is how great we are; we are now in the position of God. The technology is the same, but the apprehension of it, the moral that we draw is very different: it can be either a reminder of our created nature, or a denial of it.

ANNE MARIE REIJNEN

Yes. I might add that alongside the preference of the Protestant Reformation for the Word you will find the belief in the *claritas* of the Scriptures. The Scriptures, if you read them with all the tools, of course you have to master the Hebrew and the Greek among others, and you must read one part of Scriptures explaining it by another part of Scriptures—those are sound principles of reading a text—are supposed to be Word of God and unambiguous, whereas the image is ambiguous. Now we agree that it is much more complex than that, it is not for nothing that fundamentalism started on Protestant grounds, because it has to do with this belief in a meaning that is easily accessible and that is unambiguous, whereas the image can easily become an idol.

ELIZABETH THEOKRITOFF

And the Scriptures cannot be an idol?

ANNE MARIE REIJNEN

Literally speaking no because to hear you close your eyes. Martin Luther says "Close your eyes and hear. The Word is there to be heard and understood, not to be seen and touched."

ELIZABETH THEOKRITOFF

Yes, but can that Word not become an idol? Can one not make an audible idol just as easily as a visible one?

ANNE MARIE REIJNEN

Yes, we have seen it can become a fetish, or an idol or a mascot. It can be many things indeed.

SIMON SCHAFFER

I do not understand how *kosmos* is supposed to work as a place where Christians and non-Christians can work together. On the one hand it might be taken to be grandiose, if one needs *kosmos* with all the senses that you so brilliantly exposed for us to enable a place where all those kinds of dialogues can go on; that seems a very high demand to make on the condition of possibility of such a dialogue. On the other hand it is not yet obvious to me what's for you is new in this revivification of the *kosmos* position. Let me try to explain what I am concerned about. You splendidly remind us that this is not—to put it mildly—the first time when secularism has been worried about imminent destruction of the world, and this is not just the result of human technology, it is a Cold War image. And you showed wonderfully that this is after the Sputnik, it was about making better systems to deliver nuclear weapons. It was about investments by ex-Nazi rocket scientists and all the surveillance technology that went along with that. And we are still there. This is not a fantastically distractive moment that has suddenly disappeared from our future.

ANNE MARIE REIJNEN

I'd like to make two points. I would say we are no longer in the Cold War mentality, although in the West the fear of nuclear winter, for instance, and the fear of annihilation by the nuclear arsenal has disappeared when the objective grounds for that fear have not, but that is the way it is. I mean, today's fear is the global heating of the planet, and it doesn't matter to see which comes first but I would say we are no longer entirely there, but it is important to remember that this is the context in which this image is born. And I would even go further now and say that it is, maybe, the last stage of a colonial society, space the last frontier. As for the first point, I think it is probably more transparent to me that *kosmos* cannot be taken for granted and that it is relatively new for Protestant theologians to talk about *kosmos*, because Protestant theology is marked by a history of conflict against *kosmos*, in favor of "creation." Very briefly, I allude to *kosmos* versus creation, and Karl Barth is of course present to be counted with, you cannot entirely ignore him. Barth's project is first a theology of the Word of God: in that sense he is very reformed, he is a Calvinist. But I think the primary movement in Karl Barth is *diastasis*, is conflict. You have to overcome *kosmos* with its pagan, Hellenistic connotations, idolatry is something you see. It seems that Barth said: You mustn't say "es gibt" (there is, "il y a"), you must say: "Er gibt," God gives!

Kosmos is there ("es gibt," "il y a"). You don't have to believe in *kosmos* because the *kosmos* is there, but you must believe in creation, and creation and redemption go together. The great work of Gerhard Von Rad and his theology of the Old Testament was really to link together redemption and creation. Von Rad showed that it is really an accident that the book of Genesis is the one that opens the Bible, it is not really the logical or the theological place for it, you could start almost anywhere else, with the Sinai, for instance.

SIMON SCHAFFER

It is for exactly this reason that I still wonder why *kosmos* is for you a plausible ground of communal debate. You have made it seem, in a certain sense, even more difficult for us to accept that construction.

ANNE MARIE REIJNEN

Simon, maybe it is the Barth in you that has reacted!

SIMON SCHAFFER

Yes, I didn't know that I had an inner Barth, it is taking me quite a long time to work that out!

ERIC GEOFFROY

I just want to make sure that I understand really what we said at the beginning: for Luther hearing is the first sense and for Calvin it is seeing? If hearing is the first sense according to Luther, is there a special meaning about hearing? I ask that because it is exactly the same in Islam: hearing is the very first sense in the Koran.

ANNE MARIE REIJNEN

Yes, Paul Santmire in his history, which you find in the book by Gottlieb, makes a good point about the Medieval cathedral of reason that is an experience of "synthesis," which means that all the senses were used: seeing, smelling incense, touching the relics, you can actually touch the relics and other surfaces. Luther's emphasis on hearing was a reaction, you can see it as a "purge" of the senses and a concentration on the one thing that counts, which is to be a hearer of the Word of God, to be receptive.

ELIZABETH THEOKRITOFF

I am intrigued with this idea of the Word of God as something restricted to the hearing, because it seems to me that this is very different from the function of the Word in the fourth Gospel, at least, and as developed particularly in the Greek Fathers. The Word of God, far from being simply something that you hear, is the foundation, the agent of all creation, of everything we touch, see, and smell. "Word as something you hear" seems to be a great diminution of the scriptural concept of the Word of God.

ANNE MARIE REIJNEN

Regarding Elizabeth's question: it was my rendering that was narrowing, in accordance with some Protestant understandings of the Word that have been indeed narrowing. But the Word, if it is taken in its dynamic dimension, as I did, also includes music. For Luther, Word is sound, the Word is a sound that also comes as music, so music is extremely important. Finally, Luther follows on the track of St. Augustine when he talks about the Sacraments as "visible words."

Finally I would like to go back to the anthropic principle, which was sharply ridiculed by Pagels. He congratulates a Protestant theologian for not believing that you can find God's immanence in creation. However, the anthropic principle can be ridiculed but we cannot do without some form of anthropocentricism, that's really what I wanted to say. Ecology supposes autonomous active agents who want to do something for the good and I am sure about this, we are not earthworms, because you cannot say to earthworms "recycle your waste and be careful with your water." So, yes we have to be moderately anthropocentric, within a biocentric or cosmo-centric framework.

NOTES

1. H. Paul Santmire and John B. Cox Jr., "The world of nature according to the Protestant tradition," in *The Oxford Handbook of Religion and Ecology*, edited by Roger S. Gottlieb (Oxford: Oxford University Press, 2006), 15–146, 119.
2. Ibid., 120–122.
3. Ibid., 126.
4. Paul Tillich, *Systematic Theology*, vol. III: *Life and the Spirit* (Chicago: University of Chicago Press, 1963), 91.

8

Roman Catholic Contributions to Address the Current Ecological Crisis

Andrea Vicini

Introduction

Kiribatia ndN oah

People living in the small Republic of Kiribati—the I-Kiribati—strikingly experience one of the many consequences of the current environmental crisis: gradually and progressively, the Pacific islands in which they live will be submerged by the water level increase caused by global warming. They will be forced to relocate elsewhere. They will join the many others who will be displaced people because of environmental global changes.[1]

As Maryanne Loughry and Jane McAdam write,

> Kiribati is an island nation consisting of one island and 32 low-lying atolls (with a total land area of 811 square kilometres) in the Pacific Ocean. [...] Most of the land of Kiribati is less than three metres above sea level and on average only a few hundred metres wide. [...] Kiribati has one of the highest poverty rates in the Pacific. Kiribati is also thought to be one of the nations most vulnerable to the impact of climate change. This is due in combination to the low lying land mass with the population having no recourse to higher lands, the nation's limited sources of income, and the concentration of the majority of the population on one dominant atoll. These factors, combined with increasing changes in climate, pose a threat to Kiribati's food and water security, health and infrastructure, as well as the ability of the Kiribati government to cope with increasing climate-related disasters.[2]

How do the I-Kiribati respond to this crisis, to the prospect of abandoning their land before it will disappear under water? The government is promoting and supporting the gradual relocation of all citizens. The goal is to re-create the national community elsewhere, to avoid losing the national identity. Churches and associations joined in the effort to support the

I-Kiribati in building anew their future. Solidarity and care for those who are in need are at the core of Christian and Catholic beliefs and practices. At the same time, as believers we want to understand what is going to happen to them and why it is happening. It might surprise us to discover that, among the I-Kiribati, many turn to the Bible looking for answers. God promised to Noah that the earth will never be submerged again.[3] Hence, they trust this promise. They understand it literally. Their islands will not be underwater. God will intervene to keep that promise. God will take care of their needs. God will not abandon them.

While we stand in awe because of their faith in God's providence, we also wonder how we can help their faith to read the Bible in light of the signs of times, of what we are causing to our earth, and of what they need to do for their survival and their own well-being.[4] Nature has its own mechanisms and dynamics. Global climate transformations slowly occurred in the last centuries of human civilization. But, since the beginning of the industrial era, they are rapidly increasing. The greenhouse effect and the consequent global warming indicate that the earth's ability to adjust to the changes caused by our way of living is compromised.

Hence, the divine promise after the deluge implies and requires our commitment and our responsibility. It clearly suggests God's unconditional commitment to the good of humankind. But it also requires our creative and efficient participation. Our collaboration and contribution should avoid any threat to our living conditions, to the quality of life on our planet by preserving the conditions that will allow us continue to live on earth. In the biblical experience, the Noachic covenant indicates a turning point. From that moment on, the divine intervention is understood as unconditionally in favor of humankind. Hence, this covenant indicates that we will never experience God against humankind through nature. The natural world, and its powers, will never be used as a tool to punish us, not even when we stray from God's project of justice and love.

Moreover, besides the Noachic covenant, the whole Bible confirms God's willingness to relate to us as our Creator and Savior. God continues to create and to save throughout the history of humankind. Liberty is essential to understand God's agency in creation.[5] All living beings are entrusted with freedom. It is manifested in terms of vital dynamisms and agency. As believers, we affirm that, in the whole creation, we can recognize the divine presence through contemplation and discernment. What the environment will become and what will happen to us on earth depend on natural dynamisms and on our interaction with them. As human beings we already discovered and we continue to discover many of those dynamisms. Surely, we have influenced and modified a large number of them.

For the I-Kiribati, the anticipated, expected, and progressive loss of their land will neither be a divine betrayal of the Noachic promise, nor the vengeance of a wicked environment. It is caused by the environmental transformations that are occurring in our planet. They depend on natural mechanisms that we discovered in recent decades and that we are getting to know more and more. They are greatly amplified by our lifestyle, pollution, energy consumption, and use of resources, particularly in Western countries. The biblical promise that we will have a land where milk and honey flow—that is, the promised land—is true, but our ethical commitment and concrete choices will contribute to fulfill it.

Particular and Global

Kiribati is not an isolated case. In the Pacific, other island nations, like Tuvalu, face a similar destiny. To them, we can add the many costal areas around the globe that will be affected by water level increase. It is difficult to quantify the number of people that will be forced to relocate, with the social, economic, and political implications that this implies. We could affirm, however, that, despite the seriousness of these transformations for the millions of people involved, the raise in ocean level is solely one phenomenon. While it is problematic and burdening, we might say that it concerns only a limited amount of people. It will change our coastal topography. It will affect urban settlements on our shores, but those changes will not radically threaten our current way of living in the richest parts of our planet. Humankind will address this crisis too. Some will pay a higher prize than others, but we will recover from it and we will go on with our "business as usual." To express it in other words, it will be a *particular* problem and it will require ad hoc solutions. In some instances, we will be efficient and timely in providing the right solution. In other cases, our responses might be slow or insufficient and the most vulnerable will suffer the consequences more than others.

The claim that environmental issues are particular and that what is global takes priority over and against any particular concern, however, is strongly challenged. In the recent decades, environmental scientists studied the environment and its transformations caused by the lifestyle in industrialized Western countries. Our energy consumption, pollution, and use of world's resources are under scrutiny. Predictions concerning the consequences differ. Contrasting hypotheses and theories are discussed. Solutions to address the occurring environmental transformations are debated. By examining those debates, scientists seem to agree that environmental issues are a *global* problem. They require global solutions. What concerns a very particular case, like a small nation composed of a few islands in the middle of the Pacific, should be considered as inseparable from global environmental concerns.

To avoid simplifying dichotomies, we affirm that environmental issues, and the ethical challenges that they pose to us, are both particular and global at the same time. They transform the relation between particular and global. What is global affects us in the particularity of our situation. Moreover, at least at a certain extent, the particular choices that we make are significant— not only individually, but also politically, internationally and nationally. To have a positive impact on the environment globally, to our particular choices aimed at promoting sustainability we should add the similar choices of many others and of all humankind. The presumption is that there is still a *window of opportunity* to promote sustainability concretely and effectively. But this window of opportunity might change because we are gradually progressing toward a breaking point, a *tipping point* that will break this unity between particular and global. Today, we can still play a relevant role in protecting the environment with global strategies and policies and with our own ethical commitment to sustainability.

CATHOLIC CONTRIBUTIONS

Does our Catholic faith provide us with any helpful tool to address these concerns? Can we respond to the current environmental crisis by promoting sustainability and by reaching out to particular and global needs? The I-Kiribati turn to *Scripture* by suggesting that there we can find helpful elements not only for our life journey as individuals but as a people. They read Scripture in light of their personal and communal experience, and of their relationship with the whole creation. This is a typically Christian and Catholic approach, but we should read and interpret the Bible with critical awareness and profound discernment. Throughout history, our human *traditions*, as well as the Christian and Catholic *Tradition*, played a role too. Such a role is rooted in the Gospels' witness and proclamation through the apostles and their successors throughout history. This Tradition has shaped the Catholic contribution up to today by influencing our relationship with creation and with the whole cosmos. Hence, firstly, I briefly discuss this historical contribution. Secondly, I propose a synthetic ethical approach to address the current environmental situation.

Throughout History: A Theological Overview

> From a biblical point of view, of course, the whole point of the universe is to manifest God's glory, but for the present God's glory is revealed characteristically in a *kenosis* that endows the world with a surprising degree of autonomy. The self-emptying God refrains from overwhelming the universe with an annihilating divine presence. But in the mode of futurity God nonetheless nourishes the world constantly by offering us a range of new possibilities— such as those depicted by evolutionary science. [...] We might also say that God is more and better than a planner. A God whose very essence is to be the world's open future is not a planner or designer but an infinitely liberating source of new possibilities and new life.[6]

With these words John F. Haught attempts to describe our understanding of God's creative action within creation. It is a dynamic and a liberating action. It moves us to awe and, at the same time, it strengthens our human responsibility in taking care of creation. Pierre Teilhard de Chardin adds "that creation has never stopped. [God's] creative act is one huge continual gesture, drawn out over the totality of time. [Creation] is still going on; and incessantly even if imperceptibly, the world is constantly emerging a little farther above nothingness."[7] This also depends on "the interior vein of 'consciousness' running throughout cosmic history, and especially in the dramatic depths of life, that allows the Spirit of God to penetrate the natural world, luring it toward more intense modes of being."[8] For the French Jesuit, this intimate relationship between us and the Creator within the creation permeates the universe.

The Catholic systematic theologian Elizabeth A. Johnson suggests that, within the Christian tradition, the record on creation is mixed: we both lose and find creation.[9] The natural world is well present in the Hebrew Bible and it influenced early Christianity, as it appears in the New Testament writings, with the strong emphasis on Christ as the first born of the whole creation. In the twelfth and thirteenth centuries, the appreciation for the natural world reached its peak. Creation disappeared gradually in the post-Reformation period, probably for political reasons. Galileo is an emblematic example. The efforts of Teilhard de Chardin to connect evolutionary science with Catholic theology and, then, the Second Vatican Council's openness to the world and its willingness to interact and to dialogue, represent two recent positive attempts to recover the importance of creation. Hence, Johnson affirms that "Recovering the cosmo-centric power of the fuller Christian tradition puts us in line with our ancient and medieval forebears and fosters the intellectual and moral integrity of our discipline."[10]

To confirm Elizabeth Johnson's insights, the time of Galileo's controversy appears quite remote when we read Catholic contemporary theologians. As an example, the Australian Denis Edwards writes that:

> What the sciences show is that the universe does evolve with time, in the direction of increasing complexity that includes the emergence of stars, the appearance of the first self-replicating bacteria, and the evolution of human beings. The sciences do *not* reveal a divine design or blueprint. But the scientific evidence is open to Christian interpretation. This modest claim, that the sciences support an overall directionality in the evolution of the universe and life, fits well with the idea of a God who is achieving purposes in creation, redemption, and final fulfillment. It is congruent with a view of God who acts creatively and providentially in and through the laws of nature, in all the randomness and lawfulness that allows and enables a life-bearing universe to evolve.[11]

In particular,

> The sciences reveal a universe that is evolving at all levels, that is constituted by relationships, that has its own integrity, and that has an overall directionality. The evolutionary character of the universe is something that Christian theologians are able to embrace positively and to understand as the way God creates.[12]

Turning again to the past and, in particular, to the concrete repercussions of our theological understanding of creation in caring for earth, John Hart is quite critical. He writes:

> Catholicism and other branches of Christianity rarely reflected on earth *in se* for almost two millennia. [...] As God's creation, part of an integrated universe, earth was to be appreciated as evidence of divine creativity and divine compassion [...] Periodically, some individuals within Christianity (such as Francis of Assisi) became known for their appreciation of pristine nature and their attitude toward nonhuman species. But most (clerical and lay) Christians were preoccupied with life to come than with life on earth. Toward the end of the second millennium, the Catholic church, through the hierarchical institutional structures and leadership and through insights of thinkers in the lay community, began to promote care for earth.[13]

Hart continues his analysis by indicating that, at first, earth was solely considered "the provider of life's necessities for humankind."[14] Later, the earth's conservation became an issue, but, again, solely for instrumental value. Finally, only in recent decades, theologians articulated a growing concern for earth in light of its intrinsic value.[15]

For Hart, the development of the Catholic thought on environmental issues is marked by three stages: (1) caring for the common good, (2) concern for the creation in crisis, (3) creation concern and community commitment. While the first stage covers most of the Christian and Catholic history, the last two stages describe the developments that occurred in recent decades. Historically, the growth in awareness started in Latin America, with the request for an equitable distribution and ownership of land and of its products. Then, it continued in North America, where the Catholic Church promoted respect for the earth, for its creatures and justice in allocating the earth's resources.

Thomas A. Nairn agrees with John Hart in identifying three periods in the recent Catholic official teaching on environmental issues: (1) from Pope Leo XIII until the beginning of the Second Vatican Council, with a focus on the existing natural order and the responsibility of humankind to preserve it; (2) from the Council through the pontificate of Pope Paul VI (1963–1978), with a growing awareness of human interdependence on earth and the need to pursue the common good; (3) during the pontificate of John Paul II (1978–2005), with a great importance given to contemplation and human cocreation but, at the same time, with a greater awareness of human sinfulness and selfishness, setting the ecological question within wider concerns regarding human society.[16]

Let us now consider more in detail Hart's analysis. In reading Augustine (354–430) and Thomas Aquinas (1225–1274), Hart finds respect for creation (i.e., caring for the common good) together with a hierarchical approach—that is, humankind dominates and subdues nature because of its instrumental value. But the goods of creation need to be distributed among all people according to their needs. This approach is centuries old. It can also be traced in the Catholic social teaching that was inaugurated by Leo XIII (1878–1903) with the first social encyclical letter, *Rerum novarum* (1891).[17] Francis of Assisi (1182–1226) renewed this approach by indicating a relational new way of interacting with creation—that is, all creatures are sisters and brothers. His influence, however, was confined to spirituality and private devotion. Until recently, it did not shape theological consciousness, engagement, and Magisterial teaching.[18] The Second Vatican Council, with its *Pastoral Constitution on the Church in the Modern World* (*Gaudium et spes*), added a concern for intergenerational responsibility for the whole earth and for its goods.[19]

John Hart's historical brief overview, however, does not reflect critically on the Patristic period and on the Middle Ages. By studying texts published during these two periods, Jame Schaefer appropriates nine historical concepts that result from her selective process of inquiry. She uses them to articulate her theological contemporary approach. In particular, Schaefer explores the anthropocentric focus that dominates these two historic periods by considering critically the historical context and the authors' worldview. The list of these nine concepts points to the need for a careful assessment of those ages and of their authors. She indicates the following concepts: (1) the goodness of creation; (2) the beauty of creation, as an indicator of the Patristic and medieval aesthetic appreciation for earth; (3) the sacramentality of creation; (4) creation's praise for God, to indicate how we can experience reverence for earth because "it manifests God's presence and character, albeit dimly";[20] (5) the functional unity of creation, to suggest the need for cooperation between human beings and other species to preserve such a unity; (6) kingship, by understanding it as companionship with creation; (7) a restrained use of creation; (8) love for the neighbor, and the promotion of all virtues, because they indicate how to live in relation with all creatures; (9) love for God's creation and, in particular, the earth.[21]

From Schaefer's study emerges "the human as a virtuous co-operator"[22] within creation and with God. The influence of Aquinas' theological anthropology, hence, is quite evident.[23] Schaefer's work invites to further critical and constructive exploration of these texts from both the Patristic period and the Middle Ages to find insights that today can help us to address local and global ecological issues by shaping our consciousness, our spiritual experience, and our ethical action. Furthermore, they can become a common ground for the whole Christianity.

By continuing John Hart's historical analysis, it is only in recent decades that the official Catholic teaching began to highlight the intrinsic value of the earth and to express its concern for the critical situation of our planet, by requesting a greater care for current and future generations. This transformation started with regional statements in the United States (in Appalachia, to support the struggle of coal miners, and in the Midwest, to promote land consolidation and conservation).[24] It continued with documents of the

Conference of the American Bishops, with interventions of Pope John Paul II, and with more recent statements of the American Bishops.[25]

In particular, in their letter *Renewing the Earth*, the US Bishops affirmed that "the Christian vision of a sacramental universe—a world that discloses the Creator's presence by visible and tangible signs—can contribute to making the earth a home for the human family once again."[26] This emphasis on *sacramentality*, that is also highlighted by Schaefer during the Patristic period and the Middle Ages, is characteristically Catholic. In his major study on Catholicism, Richard P. McBrien defines the emphasis on sacramentality as one of the most relevant characteristics of Catholicism, together with mediation and community.[27] "Sacrament" indicates a reality imbued with the hidden presence of God. We can see God's presence in all creatures, in all things, as Ignatius of Loyola would say. As McBrien writes: "A sacramental perspective is one that 'sees' the divine in the human, the infinite in the finite, the spiritual in the material, the transcendent in the immanent, the eternal in the historical. For Catholicism, therefore, all reality is sacred."[28]

Two are the consequences of this emphasis on sacrament: first, "Catholicism [...] insists that grace (the divine presence) actually enters into and transforms nature (human life in its fullest context). The dichotomy between nature and grace is eliminated."[29] Second, "for Catholicism, authentic human progress and the struggle for justice and peace is an integral part of the movement toward the final reign of God."[30] While McBrien reflects on sacramentality with a focus on humankind, the US Bishops expand the centrality of this understanding to the whole creation and to all creatures.

We can affirm that this theological vision of creation, and of humankind within it, is inseparable from concrete applications and actions that aim at protecting "the common good of all people, the common good of the entire ecosystem, the common good of the whole web of life."[31] Catholic theologians too have developed their reflection, both systematically and ethically, by relying on the common good.[32] They are well aware that, to avoid using the common good as a generic and vague concept, it is indispensable, first, to define it more accurately by situating it historically and contextually over against the claims of groups and communities. Second, we should struggle in clarifying how to articulate the common good and to promote it concretely in contemporary society well beyond the limits of religious membership.[33]

In recent decades, many Catholic theologians contributed to further enrich our reflection. I mention only a few of them. Rosemary Radford Ruether emphasized a feminist perspective. She shaped the beginnings of ecofeminism and ecojustice against any patriarchal attitude toward earth and creation.[34] Leonardo Boff advocated for an ecological liberation theology, according to which "all things in nature are citizens, have rights, and deserve respect and reverence."[35] Hence, Boff expanded our understanding of the common good by rejecting any anthropocentric reductive approach: "Today, the common good is not exclusively human; it is the common good of all nature."[36] Thomas Berry's eco-theology continued to advocate against anthropocentrism. It also invited to recognize and to respect the rights of the whole biotic community, so that "every component of the Earth community would have its rights in accord with the proper mode of its being and its functional role."[37]

The current theological debate among Catholic theologians highlights further elements. First, in light of his concern for reconciliation within society and history, John Pawlikowski stresses the need for reconciliation between humanity and the rest of creation. He writes:

> A sound ecological ethics will emerge only within a theological context where God is understood as sharing with humankind responsibility for the maintenance and the development of creation, to a degree never before conceivable, and where high priority is assigned to the reconciliation of humanity with the rest of creation. Additionally, such an ecological ethic must be guided by three fundamental convictions: (1) all creation is integral to the ongoing process of salvation, leading to the emergence of the final reign of God; (2) humankind must act in a manner that insures the preservation of creation for future generations, because the passage to the final divine reign is one of transition, not destruction; and (3) through the gift of co-creatorship men and women share with God in the process of bringing the divine reign into realization.[38]

In other words, reconciliation calls for choices and for practices that make it evident in today's social fabric. Such a responsibility concerns single individuals, the Catholic Church, and society at large. We can add that parishes and religious orders have a role to play in this ongoing process. In many instances, local Catholic *parishes*, with their concrete projects, show a growing involvement in caring for creation and in protecting it. Second, increasingly, *religious congregations* are committing themselves, together with lay people, to reflect and to be more engaged in practices that promote sustainability.[39] As an example, in their last General Congregation (2008), the Jesuits affirmed that: "Care of the environment affects the quality of our relationships with God, with other human beings, and with creation itself. It touches the core of our faith in and love for God."[40] Hence, the Jesuits "should invite all people to appreciate more deeply our covenant with creation as central to right relationships with God and one another, and to act accordingly in terms of political responsibility, employment, family life, and personal lifestyle."[41] Finally, they urge "all Jesuits and all partners engaged in the same mission, particularly the universities and research centers, to promote studies and practices focusing on the causes of poverty and the question of the environment's improvement."[42] Concretely, a task force composed of Jesuits and lay people has been named and it has already started to reflect and to interact in order to make practical suggestions and to promote sustainability concretely.[43]

In conclusion, Hart's assessment of the Catholic historical contribution to sustainability is overall quite positive. He integrates his reflection focused on the official Catholic statements and on theological contributions, by paying attention to the believers' involvement on specific projects aimed at promoting justice, at advocating, and at lobbying to support these projects (e.g., in the case of water protection against attempts to privatize it), to promote rights and the common good.[44] Hart writes: "The Catholic church in its various forms has offered significant insights into environmental issues from a religious and ethical perspective. It has also sought, through actions by its official bodies and through support for projects undertaken at the grassroots and parish levels, to foster care for earth, understood as God's creation."[45] He also recognizes an ongoing transformation in the Catholic official teaching, from a more anthropocentric perspective to promoting stewardship, with an emphasis on caring for God's creation and even with "a sense of kinship with earth and all living creatures."[46] Hence, interdependence and a comprehensive understanding of relationality to include all living creatures and the whole creation characterize the Catholic contribution to promote sustainability.

We can further integrate Hart's study by briefly considering the most recent Magisterial developments. They continue and confirm the Catholic commitment to address environmental issues.

The *Compendium of the Social Doctrine of the Church* is an example of such commitment.[47] In preparing and in publishing it, the Pontifical Council for Justice and Peace was not interested in offering new teaching on social matters, but in articulating a comprehensive and ordered presentation of the current teaching. A whole chapter reflects on "Safeguarding the Environment."[48] The biblical foundation leads to recognize God's presence in history[49] and God's relationship with creation,[50] with a positive appreciation for the technological contributions,[51] even in the case of biotechnology.[52] Technological advances should facilitate a greater respect for all living creatures.[53] The goal is promoting justice and solidarity with the whole creation.

For the Pontifical Council for Justice and Peace, the relationship between creation and humankind is currently at a critical point. The environment is a "common good,"[54] but we should strengthen our responsibility for creation[55] by respecting the integrity of creation and of its cycles.[56] Furthermore, the responsible promotion of the common good entails the equitable sharing of the earth's goods[57] and new lifestyles that promote "sobriety, temperance, and self-discipline at both the individual and the social level."[58]

Recently, Pope Benedict XVI has confirmed this approach. In his 2009 encyclical letter *Caritas in veritate*,[59] the same themes surface to highlight our responsibility toward creation, the need to promote solidarity[60] and intergenerational justice[61] among peoples by redistributing equitably energy resources.[62] For the Pope, we need a new covenant between human beings and creation,[63] and virtuous choices and behavior help us to achieve it.[64]

In his 2010 message to celebrate the world day of peace, Benedict XVI affirmed that integral human development depends on our relationship with the environment. In particular, we need to promote a new "intergenerational solidarity," as Pope John Paul II advocated in his message on the same occasion addressed in 1990.[65] Hence, the protection of the environment requires a new model of development, "one which would take into consideration the meaning of the economy and its goals with an eye to correcting its malfunctions and misapplications."[66] At the same time, it demands greater sobriety and prudence in our lifestyle choices together with the promotion of charity, justice, responsibility, solidarity, subsidiarity, stewardship, and the common good in light of "the indivisible relationship between God, human beings and the whole of creation."[67] The Pope encouraged a more respectful management and use of earth resources and he stressed how "In the light of divine Revelation and in fidelity to the Church's Tradition, Christians have their own contribution to make."[68] They should protect "the natural environment in order to build a world of peace"[69] for today's humankind and for future generations.

CATHOLIC THEOLOGICAL ETHICS: NEW APPROACHES AND INSIGHTS

Reflecting on the situation lived by the I-Kiribati people, we have seen how our interdependence and our interrelatedness call for ethical responses that should influence *particular* situations and the people involved. They should also have positive *global* implications for the whole world. Particularity and universality, often separated, should not be kept asunder any longer when we address ecological issues. Many are the situations in which we should define our way of proceeding to address particular and global needs (e.g., in the case of pollution, global warming, and so on).[70]

What is the specific Roman Catholic contribution in addressing these specific cases, as well as many other cases that characterize the current environmental crisis? I answer from the standpoint of fundamental theological ethics by highlighting briefly three ethical resources that make a relevant contribution to reflect theologically on sustainability: natural law, virtues, and Scripture.

NaturalL aw

In the Catholic Tradition, *natural law* plays a relevant role.[71] Throughout history, among the many ways of interpreting it, the Thomistic understanding appears to be the most insightful, as it is indicated by the recent renewed scholarly interest in natural law.[72] For Thomas Aquinas, "natural" should neither be understood merely from a biological point of view, nor, ethically, as deontologically normative. Nature, in its givenness, is understood teleologically, in light of its end, as the fundamental basic element that allows our ethical striving for what is essential, for what is good. In light of our conscience and experience, we are left with the task of proposing specific norms because they are not intrinsic to our understanding of natural law. The current theological challenge is to articulate an understanding of natural law that, rooted in the Tradition, is able to interpret it creatively and forcefully in supporting stewardship within creation and in strengthening sustainability as a way to tend toward what is good and just.

Virtues

For Thomas Aquinas natural law was not a system on its own, separated from the whole moral life. The *virtues* were an inseparable ethical component, integral to natural law. To the question "how should we act?," virtues allow us to answer in ways that lead us to concrete decisions and practices both at the personal level and globally. The overall ethical goal of our virtuous behavior is the preservation of creation by experiencing it as a gift. Hence, within creation we flourish and promote the flourishing of all living beings in sustainable ways. Creation is identified as a personal and common good. It allows us to relate to the Good by experiencing the Divine in our ordinary lives. Creation is inseparable from Incarnation, Redemption, and from the New Creation that characterizes the eschatological times—to indicate the contemporaneity and inseparability of the key mysteries of Christian faith.

Throughout Christianity, the virtues articulate concretely our responsibility to act.[73] Justice, charity, prudence, temperance,[74] humility, and sobriety are some of the virtues that are highlighted in theological writings that address environmental issues. These virtues do not solely concern individuals, as if the path to perfection, to accountability, and to efficacy in moral action could be lived out individualistically. On the contrary, these virtues call for personal and collective transformation and change. They indicate a process, a dynamism.

Virtues help us to articulate our moral existence and actions by choosing concrete choices, attitudes, approaches, even norms, and living them.[75] Their constant presence within Christianity—both in theological reflection and in praxis—makes us wonder whether the current gravity of the ecological crisis challenges their importance in shaping behaviors and actions.

For some skeptics, virtues are not sufficiently able to avoid the current threats to our ecosystem. Such criticism is challenging. It invites us to reflect, first, on systemic solutions (i.e., investing in innovative technology to promote sustainability in efficacious and efficient ways) and, second, on the relationship between, on the one hand, these solutions and, on the other hand, individual and social commitments. Moreover, the theological reflection (probably more Protestant than Catholic) invites us to reflect on the role played by our *sin*—personal, collective, and also structural and systemic. Since the 1970s, theologians were aware that protecting the environment and using carefully natural resources were a priority,[76] but this has not been sufficient to make us act in virtuous ways by making radically virtuous choices globally. Instead of dismissing virtues by considering them as the culprit, we need to reexamine our decision-making process, our discernment, and the goals that we are aiming at, with the inner disposition to conversion, change, and transformation that characterizes Christianity throughout its history.

Scripture

The natural law informed by the virtues leads us to turn again to *Scripture*. Together with Jewish, Protestant, and Orthodox scholars and believers, we listen to what Scripture tells us about ourselves as individuals and as humanity, about our relationships, about God's relation with us and with creation. In Scripture we do not search for passages or quotes that refer to creation, nature, and the environment. Such an approach would betray an instrumental role assigned to Scripture within a predetermined theological understanding of reality. On the contrary, we aim at understanding what Scripture is telling us when we reflect ethically on life on this planet. As God's Word, Scripture engages us. It involves us in a relationship—with God, with ourselves, with other beings, with humankind, with living creatures, with nature. These relationships are not confined to here and now, but they span throughout history. The present and the *eschaton* are related.

Hence, from the Catholic point of view, Scripture frames our religious imaginary and, consequently, our faith and practices. We look at Scripture to understand who we are and who we want to become, as individuals and as communities. In this relational process of interaction a few biblical pages might play a relevant role because they participate in shaping our understanding and our practices. Among them I indicate: the creation accounts and, in particular, the seventh Day—the Sabbath;[77] Noah's story; the Genesis' stories on brotherhood; the Magnificat; Jesus' re-creation; the *eschaton* with the New Jerusalem. As an example, the Sabbath, with the commandment to rest from any work to celebrate and to enjoy, highlights a specific way to interact with all created being and to live. Moreover, the divine promise to Noah concerning the absence of any possible future deluge is still quite powerfully present in the religious imagery concerning the possible consequences of global warming, with the expected rise of water level and the flooding of shore settlements and low rise islands—as I indicated by referring to the I-Kiribati. The theological questioning is not limited to interpreting these passages. It concerns the choices and actions that are informed by reading Scripture as virtuous agents within the Christian community as well as the questions that we address to Scripture about our past, present, and future.

The scriptural emphasis on relationality surfaces in theological reflection too, both during the Patristic period and more recently. Moreover, Magisterial documents and the theological debate highlight the tension between a relationality that favors an anthropocentric approach to environmental issues and a more recent approach that explores new ways to articulate the relationship between living beings and the whole creation without reaching the extremes of biocentrism and eco-centrism. As an example, since the 1970s the *responsibility to act* is considerably emphasized in theological reflection. At the same time, both in documents and in speeches, the recent Magisterial teaching reveals a new and strong attention to the gravity of the ecological crisis. As we have seen, it asks us to face this crisis promptly and in *solidarity* with those who are less well off and who are already paying, or will pay, the consequences of any uncontrolled progress and exploitation of the environment and of its resources (i.e., energy, water, land, wood, minerals, farming conditions, etc.).

Responsibility, accountability, solidarity, and care for those in need are at the core of Christian belief and practices. They profoundly shape Catholic teaching on social and bioethical issues. Inspired by Scripture, Catholic theological ethics articulates natural law in light of the virtues and it is committed to promote a greater environmental justice with appropriate choices and actions.

Eco-theo-politics

Our reflection on the Catholic contribution throughout history should not be limited to focusing on the following: recent Magisterial texts addressing environmental concerns; theological insights articulating the relationship between humankind, creation, cosmos, and nature; approaches in theological ethics aimed at strengthening personal and collective action (i.e., natural law, virtues, and the common good); concrete examples of grassroots commitment to promote sustainability. One further element needs to be considered and added to this brief overview.

By studying history and by critically examining the present, we should consider how the relationship between the environment, theological thinking, and Catholic religious practices has been articulated. Within Christianity, Catholics have played a major role in cultural progress and in the colonization of the Americas, Africa, Asia, and Oceania. They contributed to build towns and cities, to cultivate the countryside, to map the land newly discovered, to classify plants and animals, to study the stars and the universe, to teach and to organize the educational system from kindergarten to universities, to promote the arts and cultural achievements by commissioning architectural and artistic masterpieces, to organize and to run the healthcare system. This multilayered and multifaceted approach might be called a Christian and then Catholic *eco-theo-politics*. We should study it critically, by highlighting positive contributions and by identifying dynamics that could strengthen our commitment to sustainability.

CONCLUSION: WITH HOPE

David Goodstein asks: "Is there any hope for a truly sustainable long-term future civilization?"[78] We can add: Is there a role that can and should be played by religious traditions and, in particular, by Catholics? To both questions we answer in an affirmative way. With William K. Reilly, formerly chief administrator of the US Environmental Protection Agency, we can affirm that

> The church can be instrumental in establishing a new sense of global reciprocity, an enhanced awareness that we human beings have a duty to nurture and

to sustain the planet that nurtures and sustains us. And as the community of faith continues to make strong statements that integrate knowledge and faith—statements that engage the hearth—I think that the confluence of religion and ecology will continue to gather force until it ultimately washes over and fundamentally changes our culture.[79]

DEBATE

MATTHEW ENGELKE

The choice of the image for the conference poster of Noah raises a very interesting question. Again, I think about the extent to which human activity has any impact whatsoever. This gets us back to a point that Bruno raised in relation to a certain kind of Regan-ite reading of how we should deal with the environment, which is that we have absolutely no agency: what will be will be. One perhaps perverse way of reading the promise to Noah, which you mentioned as well, is: what does it matter? So this gets us back to the key issue that the theological papers are raising for me, which is the question of how free will and human agency should be understood vis-à-vis a divine plan. I wonder if you can use the example of the South Pacific peoples to play out a few of these points.

ANDREA VICINI

An Australian colleague, Maryanne Loughry, mentioned to me the case of the I-Kiribati people. She visited the I-Kiribati. They live on very few flat islands, very close to the ocean level. Because of global warming and the consequent increase in ocean level, gradually their land is disappearing. As believers, they trust that God will not abandon them because of God's promise to Noah. God will find a way to stop the consequences of global warming. Professor Loughry was sharing with me her concern for this understanding of Noah's promise in their own lives. She was wondering how we could help them. We should empower them to understand Scripture differently, as an affirmation of God's closeness and of God's help. At the same time, Scripture should be read more critically to support any action aimed at preserving their culture and national existence. The local government is trying to displace gradually all citizens on nearby islands, in other nations. It seems to me that as Christians—Catholics, Protestants, and Orthodox—we could do more to empower these people. We should help them to become critical readers and critical believers. They should be able to challenge what is happening in today's world.

IZABELA JURASZ

My question is rather more general. Your presentation has really shown that according to catholic theology the reflection on creation belongs to moral theology. But I ask myself if we could also go elsewhere: is it possible to think about creation in terms belonging to dogmatic theology or fundamental theology? Instead of personalizing creation, defending it, making her an equal of man, couldn't we imagine another approach to creation than the one based on the moral qualities of man needed to cope with creation? Is it sufficient to remain under the umbrella of moral theology?

ANNE MARIE REIJNEN

The question addressed by Izabela to Andrea concerns the place where the reflection on the environment and Creation is done. Most of the focus of Andrea's speech relates to what is called in Roman Catholic theology, the discipline of "moral theology" (in French: *théologie morale*), which is concerned with virtues as one important aspect of moral theology. The question is: should the reflection not be opened up toward what's called fundamental theology (also called dogmatic or systematic theology)? This discipline is more concerned about questions of the divinity, of creation, and about the ways to balance the immanence and the transcendence of God.

ANDREA VICINI

As a trained theological ethicist, I prefer to reflect by articulating my reflection within the framework of theological ethics. I turned first to systematic theologians, by considering, for example, what has been proposed by John Haught and by Elizabeth Johnson. In light of their insights, I want to reflect on God and on creation by focusing on how God acts. As an example, the Australian systematic theologian Denis Edwards develops this approach in his 2010 book *How God Acts: Creation, Redemption, and Special Divine Action*. When we consider both the Bible and the Christian (and Catholic) tradition, we try to identify how God acts in the world and, at the same time, our action in the world. God and Creation appear to be inseparable. Creation does not diminish God's glory. Creation indicates how God has chosen to express and to manifest divine glory. At the same time, creation is inseparable from Redemption. We can experience the world as already saved. Creation and cosmos are already saved. But humankind is not passive: we participate in both creation and Redemption.

We could develop further our theological reflection on creation. We could focus on the Trinity by highlighting its relational character. We could also add that we are part of this relational dimension. Our action in the world depends on this relation and expresses it in today's world.

BRUNO LATOUR

It's a great summary of two thousand years tradition. My question is linked to the three talks of this morning with one that we are going to hear about Islam. We hear about Calvinists, Orthodox, Muslims, Catholic all being fully green, recently "greened," so to speak. All religions have done some sort of eco-theo-politics, whatever their creed, since they organized the space, landscaped it, so to speak, like Jesuits in Brazil, monks in medieval France, Calvinists in America. And, of course, in Venice we see quite a lot of explicit eco-theo-politics. But recognizing retrospectively that this has always been the case is not the same thing at all. Understanding religious institutions in the past having always transformed the land, is not the same thing as understanding today in what exact religious ways the land should be transformed.

ANDREA VICINI

How "green" is Catholic theology? Could we say that it has been a "green theology?"

BRUNO LATOUR

I'm worried by the recent turn toward greenness and I'd like to understand what does it make. It will make no difference in conflict because we will not be able to say, in any practical case—for the red tuna, or for all the organization of cities, or for the revamping of Venice: "Look, being religious makes this difference or that difference." We would simply say that there has always been some strong eco-theo-political effects of all religions, but without extracting from this retrospective realization any touch stone for the present.

ANDREA VICINI

First, when we study Catholic official documents we discover that many of them begin by writing: "As we have affirmed in all previous documents..." Of course, when we read that, we realize that change is taking place. The continuity with the past occurs through a discontinuity. Likewise, when we ask ourselves "How green is Catholic theology?" we answer by pointing out both to some continuity and some discontinuity: continuity in caring for the environment from an anthropocentric perspective and discontinuity because of a gradual transformation of such an emphasis.

Second, within the Catholic Church (and, probably, also within other churches) we should continue to learn how we could become more critically aware and accountable. In theological reflection we notice a progressive increase of critical thinking. Theologians are concerned about what we did in the past, for example, during the colonial period. We were mixing the novelty of the Gospel with our own culturally colonial approach, by imposing it. In the new frontier, we brought our old world and our old cultural bias. In some instances, however, we were gradually influenced by the local culture and arts. The colonial architecture gives us examples of both approaches: in recently discovered lands, churches were built to replicate European models; at the same time, we can find other churches that show a clear local influence.

Moreover, I can mention the Maya indigenous people in Chiapas (Mexico). In recent decades, the Jesuits missionaries were able to integrate and to incorporate Mayan culture in organizing the local Catholic communities. This allowed for a greater respect for the many ways in which the Maya people relate to nature, concretely and spiritually.

ANNE MARIE REIJNEN

This is a truly Catholic view on the common Christian tradition. There is in our tradition a vast treasure room of thoughts and practices that are helpful today; we need to uncover them. The Franciscan spirituality is one stream; the virtues, a way of reactualizing Aristotle, is another. I would add a third one. It is the vision put forward by Teilhard de Chardin. In 1942, if I remember well, Teilhard wrote a very interesting book called *The Place of Human Being in Nature*. And I read this book this summer in preparation for this Dialogue. In this book, *De la place de l'homme dans la nature* he is truly visionary about the Noosphere, he seems to have an intuition about the coming of what we now call the digital age with the Internet. It seems to me we might acknowledge Teilhard de Chardin as a benevolent presence from the Catholic tradition who has done much work on the interface of religion and science. My second question is about John Pawlikowski. It strikes me that in our dialogue we seem to be forgetting sometimes that as Christians, we are Jewish-Christians. And Pawlikowski, I know, is active in Chicago on the reconciliation between Judaism and Christianity. So when you called him, do you refer to that work that is very important? It is essential to our purpose to reclaim the Jewish thought of Creation. Fundamental is the assertion that the earth is the Lord's. See the many restrictions on killing animals, and the land never being our property but always being turned over, symbolically, every 50 years at least, to the Lord (the "jubilee"). Could you maybe amplify a little bit? Because I think the Jewish voice has been lacking in the dialogue.

ANDREA VICINI

My colleague John Pawlikowski worked extensively on the Holocaust. He aimed at strengthening our indebtedness to our roots, to our ancestors and to our brothers, the Jews. This surfaced in his own reflection on environmental issues. He would like to see a more collaborative effort among Catholics and Jews by turning together to the Bible. We could look back together to our history and aim at strengthening reconciliation. This dynamism would give us the strength that we need to address today's issues.

Finally, St. Francis and Teilhard de Chardin are truly insightful resources. We could say that, within our Christian tradition, recent and older, we find relevant contributions to reflect on environmental issues. Probably, we have not yet taken advantage of them as we should.

TED NORDHAUS

I was struck by a couple of things. Following up on Bruno's comment, I wonder what this sort of recent "greening" of these theological traditions is actually mobilizing. Moreover, a lot of the today conversation seems to me revolved around a sort of anthropocentric theological and biocentric theological view points. This reminded me of some of the conversation we had about artists and artisans. And today there is a similar move, from anthropocentric to biocentric views: art for its own sake and care about Creation for its own sake. There's no art in what the artisan does anymore because we see it as serving productive purposes. And there's no nature or Creation in the anthropocentric view because we see it as primarily serving human needs.

I'm particularly struck when I look at this contemporary greening of theological traditions: they seem to me dressed up in a whole set of ways about concern for Creation. But of course, it's all grounded in the sense that we're dependent upon these natural systems and the rest of Creation for our own well-being. And that much of what has motivated this greening is actually the rising level of environmental concern, which is actually quite centrally concerned (whether accurately or not) with human well-being. Whether it's apocalyptic fears of ecological collapse or just a more general concerns about the decline of a variety of ecological amenities, this kind of imagining is still driven by essentially anthropocentric interests.

ANDREA VICINI

When we examine the Catholic reflection on environmental issues, we discover that its concern for the environment is motivated by a strong understanding of what it means to be just, to love and to care, to be accountable for creation, and to be responsible when we intervene in creation. As moral agents, we are responsible. We should assume responsibility for what we have done, for what we are not doing, for what we could do and should do.

TED NORDHAUS

I guess my question is: why do we imagine that this changing affect is not still profoundly anthropocentric?

ANDREA VICINI

We should understand anthropocentrism anew. We should reinterpret it. We begin to act from our own standpoint, by considering who we are. This allows us to interact with all other human beings and creatures. Relationality is essential to understand our moral action and responsibility for creation in a renewed way. We cannot define who we are if we forget that we are involved in relationships. Theologically, we can affirm the same about the

Trinity: God is understood in relational terms. This relationality includes the Trinity, humankind, and the whole creation. We cannot understand ourselves without remembering that we are shaped by just relationships with creation. Hence, we should articulate a transformative understanding of anthropocentrism.

We are not taking care of creation alone, as isolated individuals, but as a human community. If we separate the theocentric understanding of God and the anthropocentric understanding of human beings, we go against what we find at the core of Christianity, in the Gospel, in the Bible, and in theological reflection. The Incarnation is an example of such a relational unity between what is divine and what is human. They are inseparable.

SIMON SCHAFFER

Could you say a little more about what you think is so good about the common good? I would like to make a parallel between the common good and the common place Anne Marie was talking about. The reason I would like to hear more about that—common good and common place—is that there are very good political reasons to be extremely hostile to notions of the common. It was under the sign of the common good of all people that, for example, several islands in Kiribati were made to disappear not because of global warming, but because they are made of phosphate. So, some of our Micronesian and Melanesian friends have nowhere to live now because their land has been dropped by airplanes on New Zealand and the United States in the name of the common good.

ANDREA VICINI

While I point to the common good as a possible ethical resource, we should remember the critical remark of the German theologian Dietmar Mieth. In reflecting on German history and, in particular, on the Nazi period, Mieth highlighted the risk of understanding the common good as a negative and troubling great narrative. In Germany, any reference to the common good evokes the Nazis' manipulation of our understanding of what is "good" and of the "common good" for Germany and for the world. I am aware of this risk. First, I do not want to consider the common good as a great narrative with hegemonic and violent outcomes.

I propose to understand the common good in a more insightful and dynamic way. The common good is not something that is out there, already defined and clear, that I can grasp and possess. The common good depends on a difficult search. It results from a collaborative and participatory effort, that is, what we are trying to do here during our dialogue. Hopefully, the common good does not depend solely on our own search, but it also results from what others are trying to understand and to do elsewhere. Finally, we pursue the common good together.

The common good is not a great narrative in a second sense. It is not a narrative that explains everything about anything. It is not a strategic tool to pursue a predetermined agenda. Hence, it is not justified by political reasons and political hidden decisions. The common good requires an ongoing and dynamic definition. It needs to be continuously critically reviewed. We try to understand what it might mean for humankind in general and also in particular, like in the case of the I-Kiribati people.

Finally, the common good requires that we explore the relationship between, on the one hand, my personal good (and the good of my own group) and, on the other hand, the good of humankind, of a nation, of the whole world, and of the whole creation.

NOTES

1. As an example, see: Mary M. DeLorey, "Economic and environmental displacement: implications for durable solutions," in *Driven from Home: Protecting the Rights of Forced Migrants*, edited by David Hollenbach (Washington, DC: Georgetown University Press, 2010), 231–247; Christopher Llanos, "Refugees or economic migrants: Catholic thought on the moral roots of the distinction," in *Driven from Home: Protecting the Rights of Forced Migrants*, edited by David Hollenbach (Washington, DC: Georgetown University Press, 2010), 249–269.

2. Maryanne Loughry and Jane McAdam, "Kiribati: relocation and adaptation," *Forced Migration Review* 31 (2008): 51–52, at p. 51: "The population is approximately 92,000, of whom nearly 50,000 live in South Tarawa, a highly dense area with a population growth rate of 3% per year. Most of the I-Kiribati are engaged in subsistence activities, including fishing and the growing of bananas and copra (dried coconut). The soil on the atolls is very poor and there is little opportunity for agricultural development. However, the fishing grounds are rich and copra and fish represent the bulk of production and exports."

3. See:G enesis6 –9.

4. Among the authors who reflect on Scripture in light of environmental concerns, see: Dianne Bergant, *The Earth Is the Lord's: The Bible, Ecology, and Worship*, edited by Edward Foley, American Essay in Liturgy (Collegeville, MI: Liturgical Press, 1998); Steven Bouma-Prediger, *For the Beauty of the Earth: A Christian Vision for Creation Care*, Engaging culture (Grand Rapids, MI: Baker Academic, 2001); Steven Bouma-Prediger, *For the Beauty of the Earth: A Christian Vision for Creation Care*, edited by William A. Dyrness and Robert K. Johnston, 2nd ed., Engaging culture (Grand Rapids, MI: Baker Academic, 2010), in particular, 81–110.

5. On divine agency, see: Denis Edwards, *How God Acts: Creation, Redemption, and Special Divine Action*, Theology and the sciences (Minneapolis MI: Fortress Press, 2010).

6. John F. Haught, *God after Darwin: A Theology of Evolution* (Philadelphia, MD: Westview Press, 2008), 127–128.

7. Pierre Teilhard de Chardin, *The Prayer of the Universe: Selected from Writings in Time of War*, trans. R. Hague (New York: Harper & Row, 1968), 120–121. Quoted in John F. Haught, *Christianity and Science: Toward a Theology of Nature*, Theology in global perspective (Maryknoll, NY: Orbis Books, 2007), 130.

8. John F. Haught, *Making Sense of Evolution: Darwin, God, and the Drama of Life* (Louisville, KY: Westminster John Knox Press, 2010), 145.

9. See: Elizabeth A. Johnson, "Losing and finding creation in the Christian tradition," in *Christianity and Ecology: Seeking the Well-being of Earth and Humans,* edited by Dieter T. Hessel and Rosemary Radford Ruether, Harvard University Center for the Study of World Religions—Religions of the World and Ecology (Cambridge, MA: Harvard University Press, 2000), 3–21.

10. Ibid.,1 8.

11. Edwards, *How God Acts,*1 1.

12. Ibid.

13. John Hart, "Catholicism," in *The Oxford Handbook of Religion and Ecology,* edited by Roger S. Gottlieb (Oxford and New York: Oxford University Press, 2006), 65–91, at p. 65.

14. Ibid.

15. See: ibid., 66. For a more detailed analysis of documents of the Catholic church's Magisterium, see: John Hart, *What Are They Saying about Environmental Theology?,* What are they saying about (New York: Paulist Press, 2004).

16. Thomas A. Nairn, "The Roman Catholic social tradition and the question of ecology," in *The Ecological Challenge: Ethical, Liturgical, and Spiritual Responses,* edited by Dianne Bergant, Richard N. Fragomeni, and John Pawlikowski (Collegeville, MI: Liturgical Press, 1994), 27–38.

17. Ibid.

18. Hart, "Catholicism," 69–71.

19. See: Second Vatican Council, "Pastoral constitution on the church in the modern world (*Gaudium et spes*)," in *The Documents of Vatican II,* edited by Walter M. Abbot (New York: America Press, 1966), 199–308, at § 70. See: Hart, "Catholicism,"7 4.

20. Jame Schaefer, *Theological Foundations for Environmental Ethics: Reconstructing Patristic and Medieval Concepts* (Washington, DC: Georgetown University Press, 2009), 10.

21. See: ibid., 9–10 and 17–286.

22. Ibid.,1 0.

23. On Aquinas, with a focus on grace, charity, nature, and divine friendship, see also: Willis Jenkins, *Ecologies of Grace: Environmental Ethics and Christian Theology* (Oxford and New York: Oxford University Press, 2008), 115–151.

24. See: Appalachian Catholic Bishops, *This Land Is Home to Me: A Pastoral Letter on Powerlessness in Appalachia by the Catholic Bishops of the Region,* 4th ed. (Webster Springs, WV: Catholic Committee of Appalachia, 1990). This letter was first published in 1975. See also: Midwestern Catholic Bishops, "Strangers and guests: toward community in the heartland," *Origins* 10, no. 6 (1980): 81–96. Both documents are discussed in Hart, "Catholicism," 75–77.

25. See: National Conference of Catholic Bishops, *Economic Justice for All: Pastoral Letter on Catholic Social Teaching and the U.S. Economy* (Washington, DC: US Catholic Conference, 1986); John Paul II, *The Ecological Crisis: A Common Responsibility* (Washington, DC: US Catholic Conference, 1990); US Catholic Bishops, *Renewing the Earth: An Invitation to Reflection and Action in Light of Catholic Social Teaching* (Washington, DC: US Catholic Conference, 1991).

26. US Catholic Bishops, *Renewing the Earth,* 6. Quoted in Hart, "Catholicism," 78.

27. Richard P. McBrien, *Catholicism*, new ed. (San Francisco, CA: HarperSanFrancisco, 1994), 9–14. See also: Lawrence Cunningham, *An Introduction to Catholicism* (Cambridge and New York: Cambridge University Press, 2009), 8 and 13.

28. McBrien, *Catholicism*: 9–10. For an important systematic study on "sacrament" and sacramentality, see: Louis-Marie Chauvet, *Symbole et sacrement: Une relecture sacramentelle de l'existence chrétienne*, Cogitatio fidei (Paris: Cerf, 1987). See also: Kevin W. Irwin, "The sacramentality of creation and the role of creation in liturgy and sacraments," in *And God Saw It Was Good: Catholic Theology and the Environment*, edited by Drew Christensen and Walter Grazer (Washington, DC: US Catholic Conference, 1996), 105–146.

29. McBrien, *Catholicism*, 10.

30. Ibid.

31. Appalachian Catholic Bishops, *At Home in the Web of Life* (Webster Springs, WV: Catholic Committee of Appalachia, 1990). Quoted in Hart, "Catholicism," 78.

32. As an example, see: Drew Christiansen, "Ecology and the common good: Catholic social teaching and environmental responsibility," in *And God Saw It Was Good: Catholic Theology and the Environment*, edited by Drew Christiansen and Walter Grazer (Washington, DC: United States Catholic Conference, 1996), 183–195.

33. See: ibid.; John Hart, *Sacramental Commons: Christian Ecological Ethics* (Lanham, MD: Rowman & Littlefield, 2006), 147–152. On the common good in Catholic ethics, see: David Hollenbach, *The Common Good and Christian Ethics*, New studies in Christian ethics (Cambridge, UK and New York: Cambridge University Press, 2002). For a historical account concerning the common good, I also indicate: Andrea Vicini, *Genetica umana e bene comune*, L'abside 52 (Cinisello Balsamo: San Paolo, 2008), 29–230.

34. See: Steven Bouma-Prediger, *The Greening of Theology: The Ecological Models of Rosemary Radford Ruether, Joseph Sittler, and Jürgen Moltmann*, edited by Barbara A. Holdrege, American Academy of Religion Series 91 (Atlanta, GA: Scholars Press, 1995); Rosemary Radford Ruether, "Ecofeminism: the challenge to theology," in *Christianity and Ecology: Seeking the Well-Being of Earth and Humans*, edited by Dieter T. Hessel and Rosemary Radford Ruether, Harvard University Center for the Study of World Religions—Religions of the World and Ecology (Cambridge, MA: Harvard University Press, 2000), 97–112.

35. Leonardo Boff, *Cry of the Earth, Cry of the Poor*, Ecology and justice (Maryknoll, NY: Orbis Books, 1997), 133. Quoted in Hart, "Catholicism," 81.

36. Boff, *Cry of the Earth, Cry of the Poor*, 133. Quoted in Hart, "Catholicism," 81.

37. Thomas Berry, *The Great Work: Our Way into the Future* (New York: Bell Tower, 1999), 80. Quoted in Hart, "Catholicism," 81. For a presentation of Thomas Berry's theological approach, see: Anne Marie Dalton, *A Theology for the Earth: The Contributions of Thomas Berry and Bernard Lonergan*, Religions and beliefs series 10 (Ottawa: University of Ottawa Press, 1999).

38. John T. Pawlikowski, "Theological dimension of an ecological ethic," in *The Ecological Challenge: Ethical, Liturgical, and Spiritual Responses*, edited by Dianne Bergant, Richard N. Fragomeni, and John Pawlikowski (Collegeville, MI: Liturgical Press, 1994), 39–51, at 49–50.

39. Among them: Sisters Servants of the Immaculate Heart of Mary, "Sustainability community" (2011), http://www.ihmsisters.org/www/Sustainable_Community/sustaincommunity.asp

40. Thirty-Fifth General Congregation, *The Decrees of General Congregation 35* (Washington, DC: Jesuit Conference—The Society of Jesus in the United States, 2008), Decree 3 § 32.

41. Ibid., Decree 3 § 36.

42. Ibid., Decree 3 § 35.

43. See: Society of Jesus, "Ecology and Jesuits in Communication" (2011), http://ecojesuit.com/

44. See: Hart, "Catholicism," 82–85.

45. Ibid., 82.

46. Ibid.,85 –86.

47. Pontifical Council for Justice and Peace, *Compendium of the Social Doctrine of the Church* (Washington, DC: United States Conference of Catholic Bishops, 2005).

48. Ibid., §§ 451–487.

49. Ibid.,§4 51.

50. Ibid., § 456.

51. Ibid., § 457–458.

52. Ibid., § 472–474.

53. Ibid., § 459.

54. Ibid., §§ 466 and 478.

55. See: ibid., §§ 467–468.

56. See: ibid., § 470.

57. See: ibid., §§ 481–482 and 484.

58. Ibid., § 486.

59. Benedict XVI, *Caritas in veritate* (2009), http://www.vatican.va/holy_father/benedict_xvi/encyclicals/documents/hf_ben-xvi_enc_20090629_caritas-in-veritate_en.html at §§ 48 and 51.

60. See: ibid., § 49.

61. See: ibid., § 48.

62. See: ibid., § 49.

63. See: ibid., § 50.

64. See: ibid., § 51.

65. Benedict XVI, "If you want to cultivate peace, protect creation" (2010), http://www.vatican.va/holy_father/benedict_xvi/messages/peace/documents/hf_ben-xvi_mes_20091208_xliii-world-day-peace_en.html at §§ 8 and 12. See also: John Paul II, "Peace with God the creator, peace with all of creation" (1990), http://www.vatican.va/holy_father/john_paul_ii/messages/peace/documents/hf_jp-ii_mes_19891208_xxiii-world-day-for-peace_en.html at § 10

66. Benedict XVI, "If you want to cultivate peace, protect creation."

67. Ibid., § 14.

68. Ibid.

69. Ibid.

70. As an example, see: David Goodstein, "Energy, technology and climate: running out of gas," in *Expanding Horizons in Bioethics*, edited by Arthur W. Galston and Christiana Z. Peppard (Dordrecht and Norwell, MA: Springer, 2005), 233–245; George M. Woodwell, "Science, conservation and global security," in *Expanding Horizons in Bioethics*, edited by Arthur W. Galston and Christiana Z. Peppard (Dordrecht and Norwell, MA: Springer, 2005), 221–232.

71. See: Michael S. Northcott, *The Environment and Christian Ethics*, New studies in Christian ethics (Cambridge, UK and New York: Cambridge University Press, 1996), 199–327.

72. As an example, see: Commission Théologique Internationale, "À la recherche d'une éthique universelle: Nouveau regard sur la loi naturelle" (2009), http://www.vatican.va/roman_curia/congregations/cfaith/cti_documents/rc_con _cfaith_doc_20090520_legge-naturale_fr.html; Stephen J. Pope, "Natural law and Christian ethics," in *The Cambridge Companion to Christian Ethics*, edited by Robin Gill, Cambridge companions to religion (Cambridge, UK and New York: Cambridge University Press, 2001), 77–95; Jean Porter, *Natural Law and Divine Law: Reclaiming the Tradition for Christian Ethics*, Saint Paul University series in ethics (Grand Rapids, MI, and Cambridge, UK: Novalis and Eerdmans, 1999); Jean Porter, "A tradition of civility: the natural law as a tradition of moral inquiry," *Scottish Journal of Theology* 56, no. 1 (2003): 27–48; Jean Porter, *Nature as Reason: A Thomistic Theory of Natural Law* (Grand Rapids, MI, and Cambridge, UK: Eerdmans, 2005); Patrick Riordan, "Natural law revivals: a review of recent literature," *The Heythrop Journal* 51, no. 2 (2010): 314–323.

73. See: Philip Cafaro and Ronald D. Sandler, *Environmental Virtue Ethics* (Lanham, MD: Rowman & Littlefield, 2004); Louke van Wensveen, *Dirty Virtues: The Emergence of Ecological Virtue Ethics* (Amherst, NY: Humanity Books, 2000); Louke van Wensveen, "Christian ecological virtue ethics: transforming a tradition," in *Christianity and Ecology: Seeking the Well-Being of Earth and Humans*, edited by Dieter T. Hessel and Rosemary Radford Ruether, Harvard University Center for the Study of World Religions—Religions of the World and Ecology (Cambridge, MA: Harvard University Press, 2000), 155–171; Ronald L. Sandler, *Character and Environment: A Virtue-Oriented Approach to Environmental Ethics* (New York: Columbia University Press, 2007).

74. "We need to use our understanding of ourselves as part of the environment to guide us in restraining our impact on the rest of the environment in ways that allow us for sustainability." Christine E. Gudorf and James Edward Huchingson, *Boundaries: A Casebook in Environmental Ethics*, 2nd ed. (Washington, DC: Georgetown University Press, 2010), xi.

75. "The cultivation of virtues allows and encourages us to integrate emotions, thoughts, and actions. Thus it fits with the ideal of personal wholeness that many ecologically minded people espouse. The cultivation of virtues depends on narratives, vision, and the power of examples." Wensveen, *Dirty Virtues*, 8.

76. See: Richard A. McCormick, "Toward an ethics of ecology," *Theological Studies* 32, no. 1 (1971): 97–107. See also: Drew Christiansen, "Ecology, justice, and development," *Theological Studies* 51, no. 1 (1990): 64–81.

77. See: Bouma-Prediger, *For the Beauty of the Earth: A Christian Vision for Creation Care*, 89 and also 81–110.

78. Goodstein, "Energy, technology and climate," 243.

79. William K. Reilly, "Theology and ecology: a confluence of interests," *New Theology Review*, no. 4 (1991): 15–25, at p. 25. Quoted in Pawlikowski, "Theological dimension of an ecological ethic," 50.

9

THE "COSMISM" OF ISLAM AS A POSSIBLE RESPONSE TO THE CURRENT ECOLOGICAL CRISIS

Eric Geoffroy

COSMIC PRAISE

In the Islamic view, the whole creation is endowed with life because it comes from "the Ever Living," which is a major divine Name. The fundamental unity of the universe stems from the Islamic principle of Oneness (*Tawhîd*), as does the awareness that all creatures are interdependent. We can find evidence of Islamic "cosmism" in the titles of the 114 *sura* of the Koran, which refer to all the realms of Universal Manifestation: (1) the astral (Star, Moon, Thunder, Storms, the Sundered Sky, the Zodiacal Constellations, the Sun, etc.); (2) the mineral (the Cave, Mont Sinai, Iron, etc.); (3) the vegetable (the Fig Tree—but in the Text, the tree and the ear of wheat are the favorite parables); (4) the animal (the Heifer, Cattle, Bees, Ants, the Spider, the Elephant, etc.); (5) invisible beings (Angels, *jinn*); and of course the human realm.

A celebrated passage from the Koran stresses the dignity accorded to Nature in the divine economy: "Surely We offered the trust to the heavens and the earth and the mountains, but they refused to be unfaithful to it and feared from it, and man has turned unfaithful to it; surely he is unjust, ignorant" (33:72). Koran exegetes fail to agree on the meaning of "trust" (or "deposit"). Does it refer to faith, knowledge, universal consciousness, and the responsibility in managing the planet? In any case this trust (deposit) must be given back in the best condition, clean and completely intact, since man is only a tenant of the Earth and the whole Creation will return to the original primordial Oneness. But to come back to *sura* 33:72, which I just quoted, what matters here is that far from being reality with no consciousness or simply inanimate objects, the skies, the earth, the land, and the mountains are worthy of being God's partners. In reality everything is a "sign," as the Koran often reminds us. Indeed: "Surely Allah is not ashamed to set forth any parable [such as] that of a gnat" (2:26).

All the realms are summoned to universal consciousness, for they are united in cosmic worship: "And whoever is in the heavens and the earth makes obeisance to Allah only, willingly and unwillingly, and their shadows too at morn and eve" (Koran 13:15). "Do you not see that Allah is He, Whom obeys whoever is in the heavens and whoever is in the earth, and the sun and the moon and the stars, and the mountains and the trees, and the animals" (22:18). "And whatever creature that is in the heavens and that is in the earth makes obeisance to Allah (only), and the angels (too) and they do not show pride" (16:49). "Do you not see that Allah is He Whom do glorify all those who are in the heavens and the earth, and the (very) birds with extended wings?" (24:41). "And the thunder declares His glory with His praise" (13:13).

This cosmic worship comes from *Fitra*, "pure primordial nature," whereby all creatures know God immediately because they have come from a divine world.[1] Clearly, primordial nature not only concerns Muslims, since *all* human souls have agreed to the Pact (*mîthâq*) with God in the spiritual world, before Incarnation.[2] As one of the early Muslim theologians points out, "*Fitra* is the absolute permanence of the knowledge of God" in man, who is thus devoted to worshiping God innately but often unknowingly. This adogmatic or predogmatic primordial Tradition, of which the only principle is the awareness of the Oneness of God (*Tawhîd*), is the subject of a *sura*: "Then set your face upright for religion in the right state (*hanîf*)—the nature made by Allah [*fitrat Allâh*] in which He has made men [literally 'natured' men]; there is no altering of Allah's creation; that is the right religion, but most people do not know."[3]

The Prophet was very responsive to the living universal consciousness because he experienced it every day, in his relations, for instance, with the mineral realm. He tells that the stones greeted him during the period before the first Koranic revelation, and he heard the stones invoke God. "That mountain loves us, and we love it," he said of Mount Uhud. As for the vegetable realm, there is the episode of the moaning palm trunk, which is celebrated because heard by all the onlookers: the trunk began to moan when the Prophet ceased to lean on it to turn to address his followers; the Prophet then comforted the trunk and the moaning ceased. The order was then given that the trunk was to be buried in the right and proper way, as if it were human...Animals are greatly respected in classic Islam because like man they have a soul-consciousness (*rûh*), albeit to a lesser degree than man of course, although the Koran specifies that human beings can fall to a degree of consciousness lower than that of an animal. According to Islam, all animals know pleasure and suffering—few Muslims are aware of this—and they will be judged and resurrected, naturally in their own ways. This is not so surprising, considering that God made a "revelation" to the bees (Koran 16:68). The same Arabic term is used for the revelation made to the Prophet.

"The earth, the seven heavens and their inhabitants celebrate God; there is nothing in the creation that does not proclaim His glory." Here the Sufi Ibn 'Arabi (d. 1240) started from verse 17:44 in affirming that all the realms are living and are expressed. "God can only be praised by that which is living and has a consciousness. The Prophet said in this sense that all creatures, dry or wet, are witnesses to the call of the muezzin."[4] Since "all things are living and look to their Lord," we must be respectful of all that which surrounds us.[5] Although Ibn 'Arabi clearly states that "the beasts possess skills which God has set in them [and to which man has no access],"[6] he claims that the highest quality in man is the *mineral*. In fact the mineral is totally subject to God. It is ontologically transparent: if you let a stone fall, it makes no resistance, as man might. It never cheats. Thus the *Kaaba* at Mecca, a stone cube, although empty inside, is the "House of God." As Ibn 'Arabi points out, the mineral knows God and speaks through Him. In the Koran the mountain fears God (59:21), the rock crumbles out of fear of God (2:74), and so on. This is not only a vision pertaining to Sufism: the theologian Ibn Taymiyyah (d. 1328) claims that "minerals were created to praise God in a language which no one can comprehend other than He who bestowed it upon them."[7]

We understand why, then, according to the great Sufi poet Jalâl al-Dîn Rûmî (d. 1273), man carries in himself the life of all the realms. Listen to what he has to say:

> He [Man] came first into the realm of inorganic things, and from the realm of inorganic things he passed into the vegetable realm but did not remember his previous condition. And when he passed into the animal state, he did not remember his state as a vegetable, save only for the inclination which he has towards that state, especially in the season of spring—like the inclination of babes towards their mothers: they do not know the secret of their desire for the maternal breast, or the novice's like inclination towards his spiritual master...the disciple's particular intelligence is derived from that Universal Intelligence...Then Man comes into the human state; he does not remember at all his earlier souls and he will be changed again starting from his present soul.[8]

Thus far from being in opposition to cosmic reality and Nature, the human not only incorporates the various realms of Nature, but goes beyond them by going beyond himself, according to a principle of infinite evolution across the various worlds.

The divine is thus not outside the cosmos, because *Rahma*, Mercy, "encompasses all things" (Koran 7:156). This *solidarity* between God and his creation clearly implies the same solidarity between all creatures: "the whole Creation is the family of God's family." In Sufism the experience of the 'Oneness of Being' (*wahdat al-wujûd*) impedes any division between spirit and matter—precisely that dualism from which the modern environmental crisis has arisen.

The Transformative Energies of Religions and the Ecological Crisis

As God's representative on Earth (Koran 2:30), man has a crucial responsibility in the management of the planet. Yet, the "God" of the Koran himself

is pessimistic at the prospect: in the same *sura*, the angels, although obedient
to God, question the wisdom of entrusting this mission to man: "What! wilt
Thou place in it [the Earth] such as shall make mischief in it and shed blood,
and we celebrate Thy praise and extol Thy holiness?" The verse ends, however, with God's reply to the angels: "I know what you do not know." This
is the same paradox and the same ambiguous position of man on earth as in
the *sura* quoted above: "Surely We offered the trust to the heavens and the
earth and the mountains, but they refused to be unfaithful to it and feared
from it, and man has turned unfaithful to it; surely he is unjust, ignorant"
(33:72). If we remain fixed on the axis of the horizontal, it appears that
God knows man is not capable of doing what is asked of him. At horizontal
level, therefore, there is no solution, and that is what we are experiencing at
the moment. Only a "vertical" kind of consciousness—but which does not
separate transcendence from immanence—may, in the Koranic view, bring
a solution. For the time being, however, this is beyond us. Here we find the
test of the "voluntary submission to God," the approximate translation of
the term *islâm*. There is a divine design for man that is beyond his ordinary
or current consciousness. The picture described by the Koran is thus for the
time being dramatic. We do not know what future touch the Koranic "God"
may add.

Moreover, we cannot dodge the issue of the dominion of nature entrusted
by God to man in the monotheistic religions. This has become very common
in certain quarters, and especially in Western Europe. In fact, as in other
monotheistic religions, in Islam, God gives the dominion of the creation to
man (*al-taskhîr*). See, for example, the following *sura*: "And He has made
subservient to you whatsoever is in the heavens and whatsoever is in the
earth, all, from Himself; most surely there are signs in this for a people who
reflect" (45:13). But the *taskhîr* has two complementary faces: freedom and
responsibility. What kind of "man" does God give this power to? To today's
degenerate humanity? In this case the Koran admits the angels are right:
"Corruption has appeared in the land and the sea on account of what the
hands of men have wrought" (30:41). No, we are talking about primordial
Adam, about the "completed man" (*al-insân al-kâmil*) that the *principial*
Islam would seem to aspire to fashion at the end of the cycle that we live
in and naturally with the aid of other spiritualities. In their condition of
ordinary consciousness, men, as the Sufis stress, live in a state of perpetual
distraction. This is illustrated in the above-cited verse 17:44, which ends as
follows: "The seven heavens declare His glory and the earth (too), and those
who are in them; and there is not a single thing but glorifies Him with His
praise, *but you [men] do not understand their glorification*." It would thus
appear that only human beings become aware and, regenerated in spiritual
terms, they will effectively and reasonably be able to manage the planet:
"Allah has promised to those of you who believe and do good that He will
most certainly make them rulers in the earth" (Koran 24:55).

In the Islamic vision, man occupies a broad specter: once he is "completed," he will be superior to the highest angels and archangels, but he is

inferior to the animals when he is in his most fallen state. It was in this sense that although Emir Abdelkader (d. 1883) supported the technical progress made by the Europeans in the mid-nineteenth century (he campaigned to convince the Near East peoples to open up the Suez Canal), he also warned them that the "Heavens would close up again" above them.

Out of respect for life and in line with the moral project of "pursuing the middle way" the Koran mentions economic concerns and the rejection of wastefulness: "Surely the squanderers are the fellows of the *Shaitans* [evil spirits]" (17:27); "Eat and drink and be not extravagant; surely He does not love the extravagant" (7:31). "The whole Earth is a pure temple," the Prophet used to say. The mosque is primarily a meeting place for men, but the ultimate temple is Nature, or rather the whole cosmos, since the mullahs have already raised the issue of which direction they will have to pray in once they are on the Moon or Mars...And when Muslims perform their ablutions, they should bear in mind the Prophet's words when recommending using water sparingly, even on a river bank. What is highlighted in this case is primarily ethics, a conduct to be passed on by the generations.

"No fish is caught and no bird trapped without some of the glory of God leaving this world." It is as if the Prophet were warning us that the divine protection and blessing (*baraka*) would gradually dwindle as the animal species became rarer. And that is what we are experiencing in our own age. Since the various realms are interdependent, the *baraka* flows between them. "The ant in its hole and the fish in the sea," he continues "call down grace on he who teaches good to men."

The cultural decay of so-called Muslim societies has led to the Islam faithful almost completely forgetting these matters. You only have to travel in these societies to realize how far their cultural practice is removed from essential Islamic teaching. Nowadays, the religious obsession with laws and ritual and the concern of many people to meet their immediate needs, combined with the utilitarian relationship with nature created by rampant globalization, has dimmed any universalist environmental form of consciousness in these societies. Muslim experts, thinkers, and some sections of the general public have only recently rediscovered the founding lesson and now wish to place the emphasis on the essential rather than on secondary matters.

Are there possible solutions? What do the Muslim thinkers, the *ulama* suggest? They mainly propose reintroducing an essential notion, a medieval Islamic discipline, that may be translated by the "ultimate aims of *Sharia*." This is not the *Sharia* as depicted in the media. The *Sharia* is a cosmic code of law. It is the equivalent of the Hindu term Dharma. But what is the ultimate aim of the *Sharia*? Human happiness. Five principles of respect are stated: respect for human life, religion, reason, procreation, and private property. The contemporary *ulama* propose using what is technically speaking a science to promote the protection of life in all forms. They have reinstated this science to respond to the current situation of the planet and humanity. Personally I feel that this is not enough because it is still a "horizontal" type reform. I believe that the solution must be *vertical and horizontal* and

therefore will rely on what I call the spiritual revolution, now ongoing here
and there in Islam and also outside of Islam. This spiritual revolution appears
especially in Islamic liberation theology, which borrowed from Christian
theology but gave it a much more metaphysical meaning: only the worship
of the One, the single God, will free man from all kinds of idolatry in consumerism, nationalism, politics, and even religion itself.

But let us turn to the Sufis. For many of them today, there can be no
authentic spiritual journey without environmental awareness. In the Sufi
view, the state of the planet cannot be separated from our spiritual state, and
the modern ecological crisis stems from the division of mind and matter.
We thus see brotherhoods working hands-on in this area. At Mostaganem
(Algeria) the *"Alâwiyya* brotherhood," for instance, has recently created the
Djanatu-al-Arif Foundation ("Garden of Knowing"), which describes itself
as a "Mediterranean Centre of Sustainable Development." Its current director Sheikh Khaled Bentounes often speaks out in agreement with renowned
environmentalists such as Jean-Marie Pelt and Pierre Rabhi.

"If one of you holds a sapling [palm tree] and hears the Hour [of the Last
Judgment] sound, he should rush to plant it in the ground!" I was recently
told that this wise saying of the Prophet has anticipated what is now called
sustainable development. Whatever the case, it gives hope a grounding and
suggests that life on Earth will continue, with or without today's humanity.

In several passages in the Koran we find the idea that God could easily
make humanity as we know it disappear and replace it with other forms of
life—human or otherwise. For example: "Do you not see that Allah created
the heavens and the earth with truth? If He please He will take you off and
bring a new creation" (14:19). The Islamic tradition—the Koran and the
Prophet's sayings—moreover teaches us that there is an extraterrestrial and
extra-human way, but this is little-known by Muslims.

In the *sura* 99, *The Earthquake*, we find the following significant passage:
"When the earth is shaken with her (violent [/final]) shaking, And the earth
brings forth her burdens, And man says: What has befallen her? On that day
she shall tell her news, Because your Lord had inspired [revealed to] her."[9]
Meanwhile the tsunami has raged on the northeast coast of Japan and the
Earth has continued to quake in Japan,[10] even though there is no longer any
mention of it in the Western media. Clearly, the significance of these verses
is above all eschatological. As such, the increase in earthquakes recorded by
scientists (at the worldwide Seismology Bureau at Strasbourg, for example)
leads many Muslims to hasty conclusions because they echo the current situation: the Earth would seem to be in revolt. They then link this up with the
teachings of the Prophet on the "end of the cycle" that we would seem to be
experiencing and that will inevitably be accompanied by various cosmic and
psychological upheavals. Yet, according to one leitmotif in the Koran, God
is constantly sending signs to humanity so that it can look after and manage
itself. According to Ibn 'Arabí, God permanently discloses Himself. From
an Islamic perspective, there can be no question of God abandoning his
creation (cf. Bruno Latour's idea: "Frankenstein abandoned by his creator,"

Figure 5.1). But if humanity fails to grasp this, if it does not reach the right level of awareness, the signs become trials. In this sense, the ecological crisis is just one of the symptoms of the lack of awareness in man.

DEBATE

MATTHEW ENGELKE

This time, in the Eric's presentation, we have the angels asking the question of why we are giving this to man, when we know that will lead to trouble. So, once again, this is the question of the role of human agency.

MICHAEL SHELLENBERGER

I'm just going to understand this first quote from the Koran: "Surely, we offered the trust to the heavens and the earth and the mountains but they refused to be unfaithful to it, and feared from it and man has turned unfaithful to it. Surely, he is unjust and ignorant." I don't understand what that means. Whom is the "we" referring to: humans or God? If it's God, God offered the trust to the heavens and the earth and the mountains, but they refused to be unfaithful: that means that they were faithful to the trust. It's literally very confusing. The other thing I don't understand is: "Surely he is unjust and ignorant." Now, these are two different things. In other words, if I defile my nest because I don't know how to keep it clean that's different than defiling my neighbor's nest because I am unjust and I don't want my neighbor's wealth. Is the Koran saying that this unfaithfulness is due to both ignorance and injustice or are they equating injustice with ignorance?

ERIC GEOFFROY

As for the first question (whom is the "we" referring to), "we" is indeed God. As for the second question, we would have to consult the koranic exegesis: why is the word "unjust" placed before "ignorant?" In our logic we can accept that it is through ignorance that man becomes unjust. According to the Muslim wise men, no word in the Koran is placed by chance. So I can't answer your question here; I would have to consult the commentaries. It's an extremely complex verse that echoes the verse on the *Khilâfa* in *sura* 2:30. In both cases there's ambiguity about man. According to our human logic—also according to the angels' awareness mentioned in the verse—it's incomprehensible why God chose man for the Earth, when He knows that he will shed blood, cause corruption, and so on. But at the end of the *sura*, everything is resolved when God says: "I know what you do not know." Here we find a superhuman logic belonging to all religious revelation: there is a divine design for man that is beyond us and that gives hope because in situations of acute crisis, like the current ecological crises, God may have the solutions for us, but we are still not aware of them. According to Islam, the "completed man" (*al-insân al-kâmil*) is superior to the highest angels and archangels, but man is inferior to the animals when in his most degenerate state. From the Islamic point of view, the human condition can extend in an immense spectrum of consciousness.

IZABELA JURASZ

I'd like to make a philological remark: in certain Semitic languages, ignorance and injustice are synonyms, just as sin and foolishness are synonyms in the Hebrew Bible.

MICHAEL SHELLENBERGER

It seems to me that what is at stake here is that the traditional green and Christian telling ecological problems is a consequence of human greed, a kind of injustice and cruelty or insensitivity. There is an alternative narrative, which is that ecological problems are a consequence of essential human imperfectability and ignorance. Now, I'm seeing here both and I'm trying to figure out if one term has been defined two ways or they're defining the unfaithfulness as a consequence of both. I'm trying to get at the implications for these environmental movements that you describe being supported by Soufi movements.

ERIC GEOFFROY

Rereading the Arabic terms mentioned in *sura* 2:30, we can say: man is firstly unjust because he is unjust unto himself, he is unjust in his inner world against himself; he does wrong to himself and then he goes on to do wrong to the outside world.

PHILIPP VALENTINI

I would like to go back to Michael's question concerning the unfaithfulness of heavens, mountains and earth. Heavens, mountains, and earth knew the risk that lies in accepting to be accountable for the deposit; if they had accepted to be accountable for the deposit, then they would have entered the realm of possibilities where it's possible to turn the back to God, so manifesting darkness, separateness and negation.

Man accepted the deposit because his constitution reflects all the divines names or attributes, these same attributes that form the canvas of the *cosmos*. In this sense he is God's mirror But existence doesn't belong to him so he has to give it actively back to its legitimate owner—God. This receiving existence from God and giving it back to Him constitutes the divine exchange of love between the perfect human being and God. This is how existence and darkness are transformed in light and justice and this is the way the quoted verse can be understood.

Before praying, which is the place where the divine exchange takes place, the Prophet used to say: "O God, illuminate my heart with light, and my eyes with light and my ears with light and let there be light on my right and light on my left. Let there be light above me and light below me, let there be light in front of me and light behind me. O God, make me a light."

BRUNO LATOUR

My question is perhaps a bit too simple: if all our religions are so green, how is it that we have so many ecological problems? Islam is green, Calvinists are green, Lutherans are green, Catholics are green. Great, then most of the work is done already.

ERIC GEOFFROY

I have a question for Andrea: why do you put—in the Catholic theological ethics—the scripture after the natural law and virtue?

ANDREA VICINI

In my presentation, I mentioned three resources in Catholic theological ethics to reflect on environmental issues: natural law, virtues, and Scripture. Such an order, with Scripture as the last one, does not indicate a hierarchy based on decreasing importance. All three ethical resources are relevant and important. Different emphases and priorities might be present, however.

These three resources relate to one another. In articulating a specific ethical approach, we might decide to privilege one or the other. Some (e.g., virtues) can be more easily shared within society at large. Others (e.g., natural law and Scripture) could be more problematic. We could also affirm that what is explicitly rational and virtuous (e.g., natural law and virtues) might be more largely shareable than what comes from a sacred text, an inspired text, like the Bible.

Finally, from a Catholic point of view, when we interact with others we should be able to listen to what comes from them and we should attempt to establish a common ground.

ERIC GEOFFROY

In Islam we understand the question from the Koran, because the Koran is a small world, and the cosmos is a big Koran. Al-imam 'Alī spoke on this. You know, the cosmos and the Koran act as mirrors.

NOTES

1. "Surely we are Allah's and to Him we shall surely return" (2:156).
2. Cf. Koran 7:172: "And [remember] when your Lord brought forth from the children of Adam, from their backs, their descendants, and made them bear witness

against their own souls [by asking them]: Am I not your Lord? They said: Yes! we bear witness. Lest you should say on the day of resurrection: surely we were heedless of this."

3. Koran 30:30.

4. *Al-Futûhât al-makkiyya*, I, 147.

5. *Al-Futûhât al-makkiyya*, Chapter 357.

6. *Al-Futûhât al-makkiyya*, III, 489.

7. On the *Fitra*, see his *Letters from Prison*.

8. *Masnavi*, IV, 3637ff. (English translation by R. A. Nicholson).

9. Note that in this verse God "inspires" (or "reveals to") the Earth, just as the Earth "reveals"/"tells her news" to the prophets; the Arabic terms are similar.

10. This was added later by the author, during the revision of his text in 2012.

10

ECONOMIES AND ECOLOGIES OF THE SACRED IN ZIMBABWE

Matthew Engelke

I am happy to talk about the specifics of the ethnographic study that I conducted in Zimbabwe in the 1990s, as well as the ways in which both the political and the economic crisis have unfolded since 2000 and 2001, which I document. But what I wanted to do more broadly is just to talk about some of the general issues and problems that I think are related, and that tie in very much with several of the themes that we have looked at before. I shall address conflict—which is one thing that in fact we have not really addressed thus far in much detail. And I think that is partly—if I can be provocative—because we are switching now from normative models to some actually existing histories, and of course models of how the world should be, or how humans should act, do not always get borne out in practice.

I want to begin with what is perhaps the most important question: is ecology—the ecological crisis or the environmental movement—giving to what I want to call (problematically) "traditional religions"—in the global South and the non-Western world—a language of universalism that allows them to become consonant with Christianity? In other words, does environmentalism, does the language of ecology, provide animist tribal traditions with a kind of new legitimacy in the light of what we imagine to be the ecological crisis? And I think a follow-on question from that is: does this necessitate or suggest the erasure of the colonial encounter as we scurry on to the increasingly small patch of land evoked by some participants, in the face of a real or imagined flood? Everything's "green" now, and tribal peoples have always been green and have always refused to accept the logic that had been imposed upon them by colonial era missionaries and states. What I want to emphasize here is the existence of a system of what I am calling, in a very weak sense, "eco-theologies," quite separate from what we have discussed thus far, and a system, moreover, which the colonial Evangelicals and other Protestants I have studied had a serious problem with. For white colonials in Africa it was a problem because Africans believed that there was a spiritual or divine essence in the rocks, in the trees, in the water, in the fetishes: this was a serious problem and this is part of the reason why I wanted to pose to Elizabeth a question about the role that Orthodox Christianities might have assigned to these material things. It is not unique to Protestant traditions; if you look at the history of Catholic missionaries in seventeenth-century Congo, the Capuchins went to the Congolese kingdom and were quite concerned with the existence of fetish objects, with the misattribution of divine power. "Power" is a keyword: there is a politics here, and it is a politics that we have not talked about. For me colonialism is a chapter in the history of iconoclasms, it is about who has the power to name the divine and the presence of the divine.

These traditional religions are systems of eco-theologies that are not and have never been anthropocentric (in the vein of the "perspectivism" argued by Eduardo in his Amazonians studies, where he shows that boundaries between human beings and jaguars are not at all clear).[1] The idea of the human as an individual with a "bounded" self is quite important in the Christian history, whereas these are a group of tribal "eco-theologists" who are not anthropocentric in the manner discussed so far. Probably, it is not a question of whether it is or is not anthropocentric, it simply never occurred to a traditionalist in Zimbabwe that this was a relevant question. Of course, humans were not at the centre of anything, not even as representatives of some kind of divine force.

I think the function of the anthropology is always to question the normative models with which we are dealing and that is what I am trying to do here, to play devil's advocate. One interesting detail is the fact that not all people have beliefs in creation, in the sense that not all societies have creation myths. So, in the way that anthropologists would describe it, it is not a question of being created or not created and there are at least some Amazonian people for whom this is the case. In Southern Africa there is a rich tradition of creation myths but there are areas of the world in which again creation is not the issue.

What I want to do to conclude is to focus on the economy. I have been struck in my work in Southern Africa by the frequency with which a certain story of the colonial encounter appears. Desmond Tutu tells this story: "When the white man came to Africa—so the story goes—he held the Bible in his hand and Africans held the land. The white men said to Africans: 'Let us bow our heads in prayer.' When the Africans raised their heads the white men had the land and the Africans had the Bible." So this is an exchange of sorts, Bible as gift or Bible as theft, and the fact that it is the archbishop of Cape Town saying this (one of my favorite churchmen), raises some very difficult issues about conflicting passions within the subject. One of the first things colonial missionaries always did when establishing themselves in Southern Africa was reorganizing the space, reorganizing the environment into what they envisioned to be a proper order, creating order out of chaos. And also rationalizing the use of the land—in Southern Africa, in many parts of Latin America, South America, and in North America—often meant: "Look, these Indians, these Africans, they're not doing anything with the land, they're not making productive use of it." So the way in which narratives of economic productivity get tied into narratives of mission are deeply interconnected here, and played out through an ecological reordering of space. If you go to eastern Zimbabwe, to the highlands of eastern Zimbabwe, you would think that you are in Scotland because of the trees that were planted there by Cecil John Rhodes. It is astonishing, the transformation of the landscape, to not quite Eden, but Scotland!

Now, the way this is played out in the African Christian movements, in the indigenous churches, was almost always in relation to expressing a sense of alienation through the idiom of the landscape. The group that I studied was named Masowe: Shona language *masowe* means wilderness, and there are several indigenous resonances there as well as biblical resonances, but the *masowe* is a dangerous place, *masowe* is the space of nature, as we would understand it, not the space of humans. So they have in that sense been marginalized and of course the biblical sense of the term "being lost in the wilderness" is also incredibly powerful as a narrative of liberation, but it is very important to focus on this other resonance of alienation as well. Now, the way in which this reconciliation between spirituality and space manifested itself during the period of my research in the 1990s was through the building of toilets. One of the reasons I wanted to include the section on the toilets that were built in the congregations that I studied was to highlight the shift in a sense of place. They can be read as a sense of emplacement and in terms of modernization as well. We have talked a lot about eschatology but we can also talk about scatology—this is a scatological reference as well as an eschatological reference.

I want to end with simply coming back to the theme of inner conflicts, conflicts among missionaries, and within African Christianities, of how to balance sacred with secular concerns, and I am using secular in the maybe most proper sense of the term: of this world, of this age. How do we deal with things of the other world vis-à-vis things of this world? How do we balance a kind of spiritual maturity with an economic maturity? It's never clear how to balance what can often be conflicting desires and conflicting commitments, the long-term consequences of which sometimes are not known. There is no simple narrative as to why deforestation in Zimbabwe has occurred, but it is tied up with all different kinds of economic, religious, and political factors.

DEBATE

SIMON SCHAFFER

As we know, Bruno claims that religion, in the various traditions elaborated around Christianity, is all about a radical change. Because of this feature of radical transformation, religion, in its Christian instantiation at least, presents itself as a rather plausible alternative to an ecological consciousness whose ethical and emotional drives do not seem to have enough petrol to carry us through the tasks. My question is: what are the examples of these radical transformations that count as admirable? If any?

One reason why I thought it was appropriate to raise that question now is because of the widely repeated story about the chiasmus of land and the Bible. We used to have one and not the other and now we have the other and not the one. Is that an example of the radical transformation we are thinking of when we wonder about the capacity of religion to energize projects?

ANDREA VICINI

I add an anecdote to the story that you described. It concerns aboriginal people in Australia. A friend of mine, an Italian Jesuit, lived with them for a long period. The relationship with the people developed nicely. As a biblical scholar, he was reading the Bible with them. They were stunned in discovering that the biblical stories (mostly from the Old Testament) where like *their* own stories. It was astonishing to discover that the previous missionaries did not read the Bible with them. This foreigner, who did not have anything to do with them, was sharing his own treasury. They found that such a treasury was their own treasury. They were deprived of their own land, of the Bible, and of their own stories, so similar to the biblical stories. Sadly, they cannot own back their land. At least, they are getting back their Bible and their stories, with the recognition of their wisdom that this entails.

EDUARDO VIVEIROS DE CASTRO

I just wanted to illustrate Matthew's point with an anecdote that was told to me by a friend who worked with the Shavante Indians in Brazil in the late 1970s. The Shavante were being missionized by the Salesians, they had been converted to Catholicism and then baptized. One day a Shavante man comes with his little son who had just born asking the missionary to baptize him, and the missionary says: "No, no. According to the Second Vatican Council nowadays we cannot baptize your children any longer, you should revert to your traditional way of naming, recognizing and socializing with your own people." Then the Shavante man got really angry and said: "First you took from us our religion, now you want to take from us your religion too? No. Either you baptize this child now or you're dead." Obviously he hadn't lost all of his original religion, otherwise he wouldn't have threatened the missionary with instant death. Anyway, he got what he wanted, that is a proper Christian Baptism.

I think this illustrates the dilemma Bruno raised: "to ecologise or to modernize, that is the question." Indeed, we told this people: "You should modernize yourselves," and now we're going to tell them: "You should ecologise, you must quit modernism because you were right: at the end of the day you were right and not we." That's a typical double-win situation we are actually about to impose to these people. Not that I think that they should keep on modernizing because you never get there and this is precisely what is intended. It's an interesting point about how unintended these consequences are. Sometimes I have a distressing feeling that some of these unintended consequences are unintended from the point of view of their immediate agents, as the poor Scottish missionaries, but they weren't unintended at all from the wider point of view of the wise and wealthy man.

MATTHEW ENGELKE

So what are the examples of radical transformations that count as admirable? When you get into the realm of unintended consequences, the jury is still out. So, I don't have any answer to that.

SIMON SCHAFFER

What sort of radical transformations are in play in the ethnography that you've described, such that they might be called religious?

MATTHEW ENGELKE

I think the most important one is the creation of individuals. Now, this is a contentious issue within anthropology; not everyone would agree with that, but I think, following Dumont's line about the way in which Christianity imagines a certain kind of individual, this has been incredibly powerful. And without it, the ascendancy of the discourse of individual rights would not have taken place. These are major transformations that have taken place: whether they're good or bad I don't know, and what an individual is I'm not sure either. It is very common, if you read the ethnographic literature, to find examples of recognition of the self in the Bible. Interestingly, it's almost always the Old Testament, and an incredible fascination with Judaism exists in Southern Africa, a fascination with the language and the culture of the Old Testament. Another very common thing to find is a belief that there are books of the Bible that have been hidden or kept from the local populations. The thinking is these occult books are the real sources of power that have not been revealed. It's also common in Melanesia: secret books that are not being revealed. So another kind of anxiety is the anxiety of access.

BRUNO LATOUR

Is it the same thing to say that all of these religions have had a project of deep transformation of the Earth and to say that they are green? One of the ways they transform the Earth is the colonial enterprise, the way to modernize and to reshape the "pagans," the "heathens," the "natives," by clothing them, by building their houses, and so on. So, one way is to say "ok, all of us Orthodox, Jews, Christians of different churches, we always were engaged into an eco-theology." But then there is the other question: what does it mean to be green for those religions? And that's maybe exactly the question that Simon has been rephrasing: what is the criterion, what is the touch stone that could make a difference when you reread this history retrospectively, on how to deal with present day controversies? We now have churches that have always been engaged into eco-transformation, but now they are trying to rethink their transformative past to transform it yet again, drawing the lessons of modernization. Such a set of lessons would go someway to overcome my puzzling about everyone being green nowadays. We would begin to make a difference between good and bad ways to transform or to renew the "face of the Earth."

MATTHEW ENGELKE

That's a very good question and I think, for them, a very difficult question. One of the things that really got me thinking this morning is that we've gone from theologians speaking to anthropologists speaking. This is a shift in voice, and in who is speaking for whom. Anthropologists are always representing another. Let me just say: this is a very difficult issue for them and it gets back to the question of radical transformations. Because alongside this recognition of a continuity, alongside the elevation of the importance of the landscape, is a very powerful narrative, particularly in the Pentecostal churches, of breaking with the past: in other words, saying that traditional African life is not acceptable—it is in fact of the devil in many discourses—is about breaking with your traditions and becoming Christian. This becomes a major problem in many African churches to the extent that it causes serious ruptures and riffs within families, because if you are a good Pentecostal in Zimbabwe—again there are many kind of Pentecostals—but if you are a particularly good kind of certain Pentecostal, this creates issues. Because if your uncle dies and your uncle is not a Pentecostal and has a traditionalist funeral, you should not go, because that is of the devil. You need to break with that past and this creates no end of family tensions as you might imagine. So, in a sense, the answer for charismatic and Pentecostal Christians to the question of whether to ecologize or to modernize, is actually to modernize—it's not to ecologize.

ELIZABETH THEOKRITOFF

There's a connection, I think, between the question of the transformations produced by Christianity and Bruno's previous comment that, if all religions have been green all along, what about the ways in which their communities have in fact changed their environment? In both cases, one can be fairly confident that the practice will have fallen short of the principle; and you could say the same about a whole swathe of religious teachings. The basic problem is in us: we do not live consistently according to the beliefs that we profess. This is the case in our relationship to God and in relationships with other humans, why should it be any different in relation to the environment? I am not saying that this is acceptable, but that this is the perennial problem: how to get from what we profess to actually living according to our beliefs. It seems to me that the radical transformation, in Christianity at least, is in the first instance in the person. (Not the "individual": it seems to me that the modern concept of the autonomous individual, far from being Christianity's great contribution, has precious little to do with Christianity.) The person is transformed as part of a community, the body of Christ; but this transformation is not something that you can legislate or institutionalize in the wider society. Basic Christian principles can, certainly, be built into the fabric of societies and their institutions; and this has been done so thoroughly that we frequently think of things like philanthropic institutions as representing self-evident norms. Probably the most remarkable transformative element in Christianity is the teaching of unconditional love, of the unity of all mankind: we are forbidden to divide mankind into "real people" (my race, my circle) and second-class humans (other races, personal or communal enemies, etc.). Most of what we think of as civilized norms of behavior is based on acceptance of this in principle. But since it goes against almost every instinct of our fallen nature, living up to it requires a superhuman struggle, which has to be undertaken anew in the life of each person.

Monastic life is a good example of the transformations that Christianity can produce. Not because all serious Christians have to be monastic, but because the monastery is a concentrated society where people's prime focus is living according to their faith. I am not even talking about the historical record, although there are plenty of stories from early monasticism of people showing an absolutely radical, selfless dedication to the needs of other people, as well as other creatures; there are equally shining examples in monasticism today. One can see, on a very small scale, a really transformed society in relation both to other people and to other creatures: to the surroundings, to the way the land is used, to the way that buildings are constructed. But how do you transfer this onto a large scale? That is the problem.

SIMON SCHAFFER

One is supposed to be struck by the fact that religious enterprises are extremely bad at mobilizing people, that's what you started by saying, but they're extremely good at mobilizing people, that's what you ended by saying. And I'm wondering about what the relation between those two is.

ELIZABETH THEOKRITOFF

I don't think I can generalize about "religious enterprises"; but as for Christianity, I am not sure that it is extremely bad at mobilizing people. I think it is extremely demanding—it sets a very high bar—and the gap between aspiration and achievement, between the starting point and the finishing post, has much to do with this. An important point about Christian motivation is the value attached to staying power. There is a story of a monk being asked: "What you actually do?" and he says, "We fall and get up, and fall and get up, and fall and get up." That, I think, is very important when confronting environmental problems because they are so enormous. In our lives as Christians, most of us do not seem to get very far along the path of sanctity; but very many of us do actually keep moving along the path, without getting discouraged simply because we keep falling down. To say that Christianity is "good at mobilizing people" would refer to quality rather than quantity. The existence of saints shows that transformation of the human being, the shift from self-interest to sacrificial love, is truly possible.

To go back specifically to the "How green was my theology" paradox: I quite accept that Christians have been slow to recognize the theological implications of the environmental crisis. And I do not doubt that Christian communities, like others over the centuries, have on occasion transformed their environment in damaging ways. But I do not think it is warranted to draw the conclusion that we are now seeing simply a "late greening" of theology, a trendy reinterpretation with the benefit of hindsight. The fact remains that Christianity dominated societies in the Near East and Europe, some of them with highly sophisticated cultures, for more than a thousand years but environmental destruction on the modern scale never took off. So, why not? Given the technological sophistication of the Eastern Roman Empire, is it not worth asking why its cultural legacy is one that seems "scientifically backward"—as Bruno puts it—to the contemporary West? Is it just a matter of historical accident, the result of the Empire falling to the Turks; or does it perhaps suggest different priorities, a different attitude to the nonhuman creation? The great change in use of the world coincides with an eclipse of the sacramental and cosmic vision of Christianity; "nature" may exhibit design, but it ceases to be a realm of divine action (cf. the idea that a miracle must involve setting aside natural laws). Perhaps it was this narrowing of vision that facilitated a sort of hybridization of Christianity with ideas that come out of a very different worldview:—eschatological hope for the transformation of all things is secularized into a doctrine of progress, and so on. Without this hybridization, might we have achieved some of the benefits of modernization without some of the environmental and other hazards? Possibly; I really have no idea. But the fact remains that our present environmental crisis is rooted in the waning of Christian influence, not in its heyday.

ANNE MARIE REIJNEN

I would like to come back to Matthew's presentation and make a few remarks if I may. First of all, John Calvin, while speaking about the world as "the theatre of the glory of God," is capable also of delivering "toilet sermons": this remark is an answer to your joke about toilet paper. There's nothing beneath the dignity of the public discourse of a theologian: the price of bread, full employment, and good sanitation in Geneva were deemed important by the Reformer, and so indeed with him you are constantly shaping something, moving toward a better city of people, because the city of God is not what it is about, that's a very common misapprehension about Calvin's project, it's about the community of people living together, using the law in its three usages.

The second point I want to make is the following: I was very happy to hear you, Matthew, speak of "the worth of the individual," because I think that there's a mood of self-deprecation, a mood of depression, of "we have done everything wrong." Sometimes we need to hear other people tell us: but we are happy that you (missionaries) come, and we surely would not want to go back to a purported pristine age of communion with nature. Some aspects of this life are extremely repressive of the individual, and I know it's terribly politically incorrect to say this; yet I feel in a way authorized to say it, because my African students tell me about this. A woman who is barren, or who for another reason has no children, will be buried outside the communal burial ground. Twins, as you know, can be exposed from birth. When the father dies, the older son has little choice but to go take his father's place literally. So, yes there is something about humanism, there is something about the project of Christianity that I think is worth defending.

MATTHEW ENGELKE

I wasn't saying that creation of individuals was a good thing or a bad thing, personally. I'm not sure the debate is over. But I think you are absolutely right, that some of the most vocal and enthusiastic "real Christians" are in Africa. These are people who take the legacy of their faith incredibly seriously. Just look at what's happening in the Anglican communion: where is the vibrancy of the Anglican communion? It's not in England. Now, as you'll know, it's also often a conservative energy in Africa, which personally I have difficulties with. Nevertheless, you are absolutely right, there are Africans who might say, "Thank you for my individual-hood, my humanity. Thank you for giving me a language of rights, thank you for not insisting I should be circumcised (if you're a woman)." Yes, there are many Africans and many other people who will say that a lot of good has come out of the colonial encounter.

ANNE MARIE REIJNEN

Well, I also don't quite recognize what I was saying in how you now render it, because I don't want to sound like a defender of the colonial project, because indeed many critical things need to be said about it.

MATTHEW ENGELKE

Yes, it's a mistake to conflate the colonial project with the missionary project.

ANNE MARIE REIJNEN

But it's also a mistake to separate them.

MATTHEW ENGELKE

There was often a very clear difference between state and mission activity in Southern Africa by and large, and in fact a lot of the missionaries who came from the United Kingdom were marginal within the United Kingdom. And they certainly didn't always have any state sanction. They often were at odds with the state.

TED NORDHAUS

There have been quite extraordinary benefits from the transformations we're talking about: there have been extraordinary costs, extraordinary tragedies, extraordinary consequences as well. I continually find it remarkable how difficult it is to actually acknowledge the benefits of modernization, which is—for better and worse—a transformation that these various religious passions in the colonial era and elsewhere have driven.

SIMON SCHAFFER

And you take that to be because of religious passions?

TED NORDHAUS

I would say that certainly religious passions played some significant part in the process of modernization. I would also say—and this gets to the second thing I really wanted to address—that religious passion often offered a sort of theological justifications. They justified ex post facto and instantiated a variety of social and other norms that these providential transformations (economic, social, and otherwise) are demanding of the various participants. Are our religions good or bad at mobilizing resources on behalf of this sort of ostensible "city of God," at dealing with the ecological crisis? My answer would be no. Actually, early Catholicism or certainly feudal Catholicism was really good at promoting values that were functional to the working of a feudal society. Protestantism was very good at elevating values that were functional to an early capitalist and mercantile society. This is not the formal purpose of the discourse but it is the actual functional purpose. When we look at the contemporary expressions of both what we call a secular eco-theology and the embracive green theology within existing formalized religious traditions, we see why they haven't had really done much to deal with the ecological crisis. You can also start to understand what those theological expressions or conceptualizations are actually serving.

MATTHEW ENGELKE

On one point I would probably agree with your argument. I suppose I'd rather be alive now than in 1066, I worry that modernization as you're describing it becomes a form of Reaganomics, a trickle-down theory. I don't have the figures at hand, but the life expectancy in Zimbabwe is pretty much comparable with a hundred years ago, infant mortality not much different. If you look at Sudan, Somalia, Zimbabwe, and some other pockets of the world, you need to account for the kind of unevenness of development.

TED NORDHAUS

I would account for that, but I would also observe that the solution to that would be more modernization, that the way that life expectancy in Zimbabwe goes up is through continued and hopefully successful modernization.

SIMON SCHAFFER

I would like to invite Matthew to tell us a little more about Earthkeeping Churches and green warriors.

MATTHEW ENGELKE

One of the important things to underscore is that this movement was started by Martinus Daneel, who is white Christian. He grew up in Zimbabwe, he's completely fluent in Shona, he's as native as he could possibly be. So, the "traditional" movement was started by a white man and the specific churches that I worked with did not work with his movement.

SIMON SCHAFFER

One reason why I was so fascinated by was precisely because of the juxtaposition between the projects that you evoke for us there and what you rightly remind us of in terms of the complete arboreal transformation of the eastern highlands around Umtali, so that one has at least two projects of pseudoendemic arboreal reclamation and regeneration, both run by white men, both of them Messiahs, both of them authors of hagiographies of their own life, both of them into nation building, and what we want to know is what kind of tree do they choose. Rhodes chose Scots pines, what do these green warriors plant?

TED NORDHAUS

They plant native trees.

SIMON SCHAFFER

One can start to say something extremely interesting about two different kinds of Eden. Politics, political economy, and religious passions involved matter at the botanic level. Does it matter which kind of tree is planted?

MATTHEW ENGELKE

This would be indigenous. But the staple crop in Zimbabwe is maize, which is not indigenous!

NOTE

1. See Chapter 11.

11

Economic Development, Anthropomorphism, and the Principle of Reasonable Sufficiency

Eduardo Viveiros de Castro

Anthropomorphism versus Anthropocentrism

I will talk about my own work on Amazonian cosmopolitics. Bruno asked me to play the contrarian in this meeting. Not knowing exactly to whom or what I should play the contrarian, I decided that I will be the contrarian of all, the enemy of all. This all here will not be taken to mean you all, of course, but rather the all, *le tout*, the one. In other words, monotheism and its double faced creature: humanism and naturalism. I shall be speaking for pagans. In behalf rather than on behalf of pagans. I have no mandate from them, of course. But the nice thing about paganism or radical polytheism, understood in a certain sense, is that precisely you do not need to abide by any authority or scripture. The pagans I shall be speaking for do not have a scripture, they do not even have a script. They are illiterate. Their technological imagination and their theological imagination were applied to all the objects and all the aims. I will start, before getting into the present Indians, with two anecdotes concerning savages from the new world visiting the great cities of Europe. One dates from the sixteenth century and is to be found in Montaigne. The other dates from yesterday and concerns my visit to Venice.

In his famous essay *Des cannibales*, Montaigne describes the travel impressions of three Tupinambá Indians from Brazil. They were invited to come and see Paris, the King and the wonders of French civilization. Let us recall that at the beginning of the sixteenth century the Tupinambá were very likely a much more well fed and a healthy population than the French or the Europeans in general. They were certainly technologically less advanced but I am ready to bet they had a quality of life that was a few orders of magnitude better than the average European citizen of the times. And yet they were in the end destroyed by this unhappier and unhealthier European social organism and machinery. After having seen a lot of Parisian *vedute* and being introduced to the King, they were asked through an interpreter what had most impressed them in the whole visit. And they said it had been three things: (1) Why so many big strong armed men suffered themselves to be bossed around by a child (meaning probably the *Dauphin*, the child King)? (2) Why that people let their other halves? Montaigne explains that the Tupinambá called the fellow men their moieties or other halves. Why did they leave their other halves to starve and die on the streets like that? (3) And why, finally, this starving half did not come up in arms and kill the other half so as to get a little of their food and riches? As you see then, what most impressed Tupinambá was not the material technology of the French but their lack of human technology, so to speak. This revealed itself to be a permanent motif of puzzlement to the Indians up to today: The violent disjuncture between Western culture (its technology tout-court) and Western society (the quality of its human relations). And the Indian problem has always been how to get hold of one without having to accept or swallow the other. Missionaries were critical elements in this complex situation.

The second anecdote refers to my visit to Venice. I confess to have been a little disturbed by the care dedicated to that fantastic feat of engineering: the Moses system. I am not being ironic here; it is a fantastic fit of engineering. While at the same time nobody seems to care of the tsunami of people washing over the city every summer, which is probably creating not a few problems for the native owner—the indigenous Venetians—as well as for the environment. I mean the stones, monuments, transportation systems, and sewages systems seem not to have found yet their own Moses system. I have been told that Venice lost one-third of its indigenous population in the past 15 years. This should be a matter of some concern to us. I am not certain what are the lessons to be drawn from this disconnection. There is an asymmetry between the care that is being taken to prevent the literal flood and the lack of care being taken to prevent the human flood that goes over Venice every summer.

But let us move to my Amazonian Indians in whose behalf I am supposed to be speaking here. Amazonian—or American, in the proper sense of the term—Indians in general have never been modern, also in the sense of having never been in nature and therefore never have been out of it. They have not transcended "nature," as the special species that has that little, but whole crucial, *supplément d'âme* that distinguishes humans from the rest of creation. Indians form a humanity that does not define itself through its confrontation with nature. Such humanity is not the same humanity because if you have no nature then you have no humanity, in the same sense either. If that concept of the nonhuman is different from ours, the concept of the human will necessarily also be different from ours. We are not speaking of the same humanity with another nature. The two halves of the cosmo-anthropological equation must be changed at the same time: other technology, other theology; other physics, other metaphysics; other continuities, other discontinuities. What we would call nature is for Amazonian Indians an array of human like societies, alien to human beings as defined in the

West. Each living species and often many of the categories of objects, be they meteorological phenomena, celestial bodies, geographic features, artifacts, each kind of thing is human, is considered human in the sense of having an inner invisible side, which is its soul. So, each kind of being is a hybrid creature, to employ an expression used by Elizabeth. Just as man is, for us, soul, every being in Amazonian cosmos is a hybrid creature because it has a soul or, as we would say nowadays, because it has a cognitive apparatus similar to ours endowed with consciousness, intentionality, imagination, and so on. Most importantly, every being or every species of being sees itself, its own species as human. The soul is always humanoid and the soul is the side of beings that each being sees when it looks at itself, as it were when it looks at its fellow beings. The shape of a vegetable, as the shape of every being in the cosmos, is only a garment of skin, to use another term evoked here. A garment of skin that hides a humanoid inner form only visible to the eyes of conspecific or transpecific specialist like the shamans who are capable of shifting their species-specific point of view and see the world as other species see it, including how they see themselves and us. When a jaguar looks at another jaguar he sees a man, and so on. On the other hand, when a jaguar looks at us, he does not see a human being at all, but a wild pig, since jaguars, as every man, love to eat wild pig meat. That is why they kill and eat humans. Wild pigs, on the other hand, see us humans as jaguars or as cannibal spirit, because we kill and eat them. So, every being in the Amazonian cosmos is human or rather sees itself as human, including us, but does not see other species as human. In some, humanity is universal but at the same time is radically local.

So we can see that in the eyes of Amerindians—and eyes are a very important aspect of the whole thing because Indians describe the differences of perception in terms of the different eyes that each species has—everything is human, but everything is not human and cannot be human at the same time. A jaguar and human beings cannot be human to each other, at least not at same time, in the same respect, and in the same relation, as Aristotle would put it. But they are human to themselves and humans know it. That is why every interspecies interaction in the Amazon is an international affair, a diplomat talk or a warfare operation that must be conducted with the utmost care. When an Indian deals with an animal or a tree he knows he is dealing with a being that is human in its own department and this hidden humanity must be radically taken into account. Let us not forget that the Indians, like any other human being, like every other animal being, must eat or otherwise destroy all the species in order to survive. They know that to live is to leave a human footprint. The difference is that the soil that they tread is also human, meaning that the footprints are left on the body of other human beings because everything is human.

The basic tenet of Amerindian cosmopolitics is therefore what I would call anthropomorphism. We could call it anthropomorphic principle to contrast it with the anthropic principle as much as possible because anthropomorphism is, to my mind, the very opposite of anthropocentrism and, in my opinion, the alternative to it. We either get anthropocentrism or anthropomorphism.

There is no third solution. The idea of a single species lost in the cosmos is literally unbearable. And the difference between anthropomorphism and anthropocentrism is very simple. Anthropomorphism means that if everything is human, we are not a special species after all. There is nothing very much special about us because everything is like us. While anthropocentrism is quite the contrary, makes humans a separate species that has something else. Anthropomorphism as a principle is somewhat similar to the idea that Anne Marie Reijnen exposed to us before, of men being central but not privileged because centrality is a perspective not a property. Centrality is a position not an essence. Humans are not animals like the others because it is animals that are human like us. The common original condition of humanity and animality is not animality as in our popular evolutionary mythology, but humanity. Animals are ex-humans, not humans ex-animals and this former humanity of animals, in Amazonian thought, is never entirely lost just like our own animal nature is never completely lost, never completely controlled by our cultural garments, and is constantly threatening to come out violently in foot and claw. In dreams, in hallucinogenic induced visions, when one is ill, in shamanic trances, the human invisible side of animals comes to the surface of things as it were; this is the common condition of all beings, the universal anthropomorphic background of being, which is spirit or soul. This is not an ecstatic experience in the sense of being a blissful, wished for experience of transspecies communion. The Indians are not tree-hugging-people at all. If you live in a human saturated universe you have to be extremely cautious. The principle of precaution is deeply embedded in the Amerindian cultural software. When you approach any element of the environment you are approaching another human, so you have to be very cautious because, as we all know, the only dangerous thing in the universe is man. Indians know this perfectly well.

What about creation and fall? Amerindian myths see the origin of culture in society, that is, they see material and political technology—cooking fire, cultivating plants, social rules, and so on—as intrinsically connected to the origins of short life, in other words of mortality. The origin of culture is the origin of death. People die because they have culture. This is in the myth. This situation came about in myth not through any sort of sin committed by humans against a divinity but rather because of a mistake. Humans made the wrong choice when offered an alternative by a demiurge to choose between the call of the hardwood or of the softwood. Obviously, the Indians responded to the call of softwood. That is why we die early instead of being tough like hardwood trees. Humans made the wrong choice and ended up no longer able to live indefinitely by changing their skins as they grow bald, like snakes do, like some insects and some trees and like women, by the way, because Indians consider menstruation to be an internal change of skin. That is why women live longer than man. Now we know. So it was a bit of carelessness, a momentary lapse of reason, not a willful act of disobedience. And though we have lost long life this was not an entirely negative outcome. If people did not die there will be no space on this Earth to put and feed the future generations, say the Indians. How would we have children? They ask.

If Indians do have a cultural aim that is central in that of having children, constituting bodies of kin because people live in others, in their kin, with their kin. So, Indians are Malthusians after a fashion. They rather had the population remaining stable then increasing production or productivity, so as to make room for more people, more needs, and more worries. Immanence, I said, not bliss. The original state of creation, the hidden moment of indigenous cosmology is one in which humans and animals were able to communicate freely with one another. More than that, this was a moment in which beings were as it were, transparent to one another. The body did not hide the soul. Any being in mythical time could be approached as being at the same time the human being it was and the animal it was about to be. Amerindian myths do not describe speciation, how animals lost their human side or how the human side move to the background and assumed, afterwards, a species-specific shape, a body precisely. Human is the original stuff nature is made of and this, to my mind, is the essence of Amerindian cosmopolitics or metaphysics or theology, the idea that the very stuff of nature is human. Regarding what we call nature, they would say, "Well, this is all made of human stuff." In line with this, there is a lovely Kashinawa myth that starts like this: "In the beginning there was nothing, just people." So, exactly the opposite of what we say: "In the beginning there was everything, everything was made and then people." I think this is an apt parable for the width of the difference that separates us from them.

ECONOMIC DEVELOPMENT AND COSMOPOLITICAL REINVOLVEMENT: THE PRINCIPLE OF REASONABLE SUFFICIENCY[1]

"Whoever comes after will have to make do"

("An old Brazilian proverb" in Warren Dean,

With Broadax and Firebrand: The Destruction of the Brazilian Atlantic Forest)

Geopoliticsa ndC osmopolitics

As the symbol that the *physis* chose as one of its guises around past century's end, Amazonia has now become an arena in which a decisive match is being played: the players involved, bringing together the micro- and macropolitical in unprecedented ways, compete over the meaning of the future. Leaving behind the dialectics of State and Nature, these two imaginary totalities that have been reciprocally constituted by a confrontation from which people and their myriad associations were always excluded (either represented by the first, or identified to the second), a new geopolitics now takes over. Exchanging the naturalization of politics for the politicization of nature, directly connecting the land to the Earth, thereby skipping over the old national territorializations, the geopolitics of environmentalism refuses to entrust the State with the guardianship of the infinite and the monopoly on totalization. Along with the State, Nature—a certain idea of Nature—must be brought down as well. Geopolitics transmutes into cosmopolitics.

We could view things, of course, the other way around, seeing the old in the new. Environmentalist discourse may be read as the cosmology of late capitalism, a resacralization of history and geography that would close the cycle opened by the expansion of the West in the fifteenth century; a dramatic reterritorialization on a planetary scale of all those local, national, and continental deterritorializations that defined world history in the past centuries: the revenge of Totality. Environmentalism would thus mark the advent of a postenlightenment Dark Age: leaving the space-time of the relations between society and supernature, the discourse of finitude and transcendence would now be articulated in the confrontation between society and nature. The Amazon rainforest would occupy, no longer merely allegorically, the place of the gothic cathedral: the "sacred canopy" can now be admired on Google Earth, the *Hylea amazonica* (matter at its most luxuriant) would take on the austere shape of the Spirit. And Society, which not very long ago was the model of all order and of any Whole, would now see itself as the very idea of disorder, as the suicidal hubris that can only redeem itself if it accepts its subordination to an organismic totality that encompasses and determines it.

So perhaps environmentalism can be taken as a kind of repetition of Christianity—as both subverting and reinvesting, in the name of more total totalities and more concrete universals, the imperial abstractions of our modern Romes—with Brazilians, incidentally, in the ambiguous role of barbarians to be converted by the missionaries of this neogospel of the middle classes (a naturist replay of the old Protestant ethic); barbarians, on top of that, entrusted with the Amazonian Grail, inadvertent warrantors of planetary salvation.

Perhaps this is all true; but environmentalism can also be seen as a radically new discourse, which refuses some of the founding partitions and basic categories of so-called Western rationality. In particular, it rejects the idea that *Homo sapiens* is the species-elect of the universe—by divine gift or historical (evolutionary) conquest—exclusively entitled to the condition of subject and agent before a nature seen as object and patient, as the inert target of a Promethean praxis. It problematizes the theologico-philosophical concept of "production" as the last avatar of transcendence—the idea that human produce against nonhumans, in an infinite movement of spiritualization that opposes matter (production as separation from nature). In exchange, environmentalism perhaps proposes an internalization of nature, a new immanence and a new materialism—a conviction that nature cannot be the name of what is "out there" because there is neither outside, nor inside.

If we understand nature that way, as a certain idea of the real, then nature designates the absolute limit of history. This is the predicament of our era: the Ecumene has been saturated by the human, culture has become coextensive with nature, ecology, and anthropology today coincide. As a reaction against the enclosing of the planetary commons, environmentalism imposes

a drastic revision of the paradigms of unending progress and perpetual development, which continue to guide our economic doctrines and ideological
pipe-dreams. Our linear and cumulative conception of history—structurally
blind to structure, to systemic circularities and reverse causalities—took too
long to wake up to the fact that misery, hunger, and injustice are not the
result of the still partial and incomplete character of the march of progress,
but one of its necessary by-products, which increase as the march continues
to move in the same direction. The Third and Fourth Worlds already are,
because they always were, part of the First World, and they are everywhere.
We went through the twentieth century with the mind of the nineteenth
century; the future shock promises to be hard for everyone.

Social and Environmental Diversity

The diversity of the forms of life on earth is consubstantial to life as a form
of matter. This diversity is the very movement of life as information, a
form-taking process that interiorizes difference—the variations of potential
existing in a universe constituted by the heterogeneous distribution of matter-
energy—to produce more difference, that is, more information. Life, in this
sense, is an exponentialization, a redoubling or multiplication of difference.
This applies equally to human life. The diversity of ways of human life is a
diversity in the ways of relating to life in general, and to the innumerable singular life forms that occupy (inform) all of the possible niches of this world.
Human diversity, social and cultural, is a manifestation of environmental,
or natural, diversity—it constitutes us as a singular life form, being our own
mode of interiorizing "external" (environmental) diversity, therefore repro-
ducing it. For this reason the present environmental crisis is, for humans, a
cultural crisis, a crisis of diversity, and a threat to human life.

The crisis sets in as soon as we lose sight of the relative, reversible, and
recursive character of the distinction between "environment" and "society."
Paul Valéry stated in the somber aftermath of the First World War that "we,
European Civilizations, now know that we are mortal." In this somewhat
crepuscular beginning of the present century, we have come to know that,
beyond mortal, "our civilizations" are lethal, and lethal not only for us,
but for an incalculable number of living species—including our own. We,
modern humans, children of the mortal civilizations of Valéry, appear to
have forgotten that we belong to life, and not the contrary. Once upon a
time, we used to know this. A few other remaining civilizations appear to
know this still. Many more, some of which we have already killed, knew this
only too well. But today, it has begun to be glaringly obvious even for "us"
that it is in the supreme and urgent interest of the human species to abandon an anthropocentric perspective. If the demand seems paradoxical, that
is because it certainly is; such is our present condition. But not all paradox
implies an impossibility; the paths that our civilization has taken have not
been at all necessary, from the point of view of the human species. It is possible to change direction, even though this means changing much of what

many people would consider to be the very essence of our civilization. Our curious way of saying "us," for example, excluding ourselves from the "environment," would have to change, for a start.

What we call environment is a society of societies, what we call society is an environment of environments. What is "environment" for one society will be "society" for another environment and so forth. Ecology is sociology, and vice versa. As the great ecologist Gabriel Tarde said, "every thing is a society, every phenomenon is a social fact." All diversity is both a social and environmental fact; it is impossible to separate them without plunging headlong into the chasm thus opened and destroying the very conditions of our existence.

Diversity is, therefore, a superior value for life. Life lives off of difference; every time that a difference disappears, there is death. "To exist is to differ," continued Tarde; "it is diversity, not unity, which is at the heart of things." In this way, it is the very idea of value, the value of all value, so to speak—the heart of reality—which supposes and affirms diversity.

It is true that the death of some is the life of others and that, in this sense, the differences that form the irreducible condition of the world never annul themselves really, they merely change place (the principle of the conservation of energy). But not all places are equally good for us, humans. Not all places have the same value. (Ecological discourse is nothing but this: the evaluation of place.) Socio-environmental diversity is the condition of a rich life, a life capable of articulating the most number of significant differences. Life, value, and meaning, finally, are the three names, or effects, of difference.

To speak of socio-environmental diversity is not to merely affirm a truth, but is a call to arms. It is not about celebrating or lamenting a foregone diversity, residually maintained or irretrievably lost—an already-differentiated difference, static, taxonomized, and taxidermized difference, packaged into separated identities ready for consumption. We know how socio-environmental diversity, taken as mere variety in the world, can be used to substitute mock differences for true ones—narcissistic distinctions that repeat to infinity the apathetic identity of consumers, who become ever more similar the more they imagine themselves to "be different."

But the arrow of real diversity points to the future, to a differentiating difference, to a becoming that goes beyond the plural (a simple variety subsumed by some superior unity) toward the multiple (a complex variation that resists totalization). Socio-environmental diversity is to be produced, promoted, favored. It is not a question of preservation, but of perseverance. It is not a problem of technological control, but of political self-determination, and metaphysical creativity.

Economic Development and Protection of Diversity

These days Brazil appears to be haunted by heady dreams of imminent grandeur. Against such millenarianism disseminated in my country—"our turn has arrived!" (our turn for what, exactly?)—I am convinced that it is urgent, not to "stop to think," but to think so as not to stop; it is urgent to begin to think carefully so as not to stop altogether. We need to learn to degrow so we do not decay, rot, choke on our own filth. Brazil is big, as a local saying goes; yes, Brazil is big indeed—but it is a small world. In the dawn of this century the Earth is not at its best. The global patterns of production, distribution, and consumption of energy by our species are downright suicidal. My country is one of the few on this little planet of ours that still has full viability from the point of view of its resource base. Brazil boasts one of the most historically and culturally most diversified populations in the world: 220 indigenous groups, an immense number of African descendants, of European, Asian, and Near Eastern immigrants; rural and urban people of the most different ethnic and cultural origins living in a variety of natural formations that, in turn, are home to the richest biodiversity of the planet. Sociodiversity and biodiversity should be our major assets. But here we are, as always, insisting on sawing off the very branch we sit on, with policies of international trade that apply a model of development that is ecologically predatory, economically concentrating, socially impoverishing, and culturally alienating. We have devastated more than half of our country in the belief that it was necessary to forsake nature to embrace history; now look at how history, with its historical fondness for irony, demands we make that very nature our passport. I am afraid we will be found wanting.

The Space of Terrestrial Immanence

Contrary to what the Extraordinary Minister of Strategic Affairs, Roberto Mangabeira Unger, said in a recent interview, Amazonia is not a "collection of trees."[2] Collections of trees exist in botanical gardens or in the estates of the superrich. Amazonia is an ecosystem, a multiplicity composed of very conspicuous trees[3] as well as an infinity of other living species—including human beings, who have been there for at least 15,000 years.

Amazonia was never devoid of people before the European invasion; on the contrary, its demographic nadir was reached after the invasion, with its epidemics, its methodical massacres, and with its forced "descents" of native populations for concentration around mission stations and commercial outposts. And the indigenous populations have found solutions of "sustainability," throughout these millennium of coadaptation with the Amazonian ecosystem (or ecosystems, as Amazonia has many, not only one), that are infinitely superior to the brutal methods of deforestation with herbicides, chainsaws, tractors, and other high-tech feats of jaw-dropping stupidity.

The Amazonian rainforest was always densely populated, and has been "ecologically managed" for quite a while; the majority of the useful species of the forest owe their dissemination to indigenous land-using techniques: Amazonia is a cultural forest, an anthropic environment. So it was never, or at least has not been for many centuries, more likely millennia, anything resembling "virgin forest." However, it does not follow from the fact that the forest is no longer virgin that it is legitimate to rape it. Or does it? Consider the analogies.

Amazonia is suffering a violent process of aggression. I say Amazonia, not the so-called collection of trees; Amazonia in its entirety is suffering, its traditional human populations as much as their myriad of nonhuman living species. Rather than simplemindedly imitating the northern European, modernist religion of "economic development," an alternative model that puts the largest forest in the world at the center of the equation is called for, since we have arrived at a moment in the history of the planet in which life is the value that is in crisis—both human and nonhuman life. It is no longer possible to do politics without considering the space in which all real politics unfolds, the space of terrestrial immanence.

I use the word immanence deliberately here. Minister Mangabeira Unger said in a recent interview that the destiny of Man "is to be grand, divine; it is not to be a child imprisoned in a green paradise"; and that "all people [*pessoas*, persons] are spirits who strive to transcend." Well, Amazonian Indians would agree with the minister that all people are spirits; perhaps they would not agree with the idea that only human beings are people or persons, but that is another problem. Certainly, however, they would not agree with the idea that all people "strive to transcend" in order to be "divine." This is an affirmation that would sound, to indigenous ears, distressingly similar to that which they have been hearing during the five centuries since the arrival of Europeans—the affirmation that they are children who need to bow to the divine message of decaying matter: rather transcendence in order to become full human beings, to wit, to be good Christians and model citizens (i.e., with plenty of faith and no land at all).

Indians are not "imprisoned in a green paradise" as the minister said. Amazonia is not a paradise; on the contrary, it is a laborious coadaptive construction, a system in dynamic equilibrium produced by the synergistic interaction of human (indigenous) technical ingenuity and the sui generis ingenuities of the sundry organisms that live there.

Indians are not imprisoned there. The idea that indigenous populations need to be "liberated," which Mangabeira Unger expounded in another, more recent text, seems to me to be grotesquely mistaken. Those indigenous groups who suffer from depression, suicide, alcoholism, as the minister laments, are precisely those who have been dispossessed of their lands—the Guarani of Southern Mato Grosso, for example—not Amazonian groups like the Yanomami, a strong and happy people, precisely for having land that fits their vital and spiritual needs. The indigenous areas of Amazonia are the least deforested areas of the whole region; and they are the essential pieces in the process of juridical stabilization of the chaotic land distribution that made Amazonia into the paradise of illegal land appropriation, political assassination, drug trafficking, international smuggling, and government-subsidized corruption. And what does the ministry propose? A "program that will regularize landownership" that is nothing but a repeat of the loathsome principle of *Uti possidetis*: the legalization of the private appropriation of public lands, which the rich and the powerful originally took with brute force.

Naturally, Indians suffer with various problems, many of them caused by the incompetence and/or corruption of the state agencies that should enforce the respect of their constitutional rights. But it also cannot be denied that the Indians have suffered other difficulties adapting to the socioeconomic (and spiritual) forms of Brazilian national society because they have chosen from very early on in history a civilizational route that is radically distinct to our own—and, which can be called the path of immanence as opposed to the path of transcendence.

Indigenous cultures are not founded on the principle that the essence of the human condition is desire, need, and lack. Their mode of life, their "system," in the most radical sense, is other. Indians are the masters of immanence. What transcendence do we have, proud Brazilians, the self-appointed representatives of Reason and Modernity, to offer them? It is more probable that the Indians will liberate us than that we will liberate them. At least in spirit.

A Pragmatic of Reasonable Sufficiency

The problem, in sum, is that of finding an alternative way of life, because there is no alternative to life. To change the life we live—to change the way of life; to change the "system." Capitalism is a politico-religious system whose principle consists in taking from people what they have and to make them want what they do not have—always and ever. Another name for this principle is "economic development." With that we arrive right in the thick of the theology of the Fall, of the infinite insatiability of human desire before the finite means of satisfying them (see Sahlins, "The sadness of sweetness"). The economists are the priests and the theologians of our age. It is not by accident that Marx spoke of the metaphysical subtleties and of the theological astuteness involved in the concept of the commodity. But it is precisely this theology of development that we can no longer accept; we can no longer accept the equation between development and economic "growth." The world of economics is paying renewed attention to the theses of N. Georgescu-Roegen on degrowth, the thermodynamic costs of the economy, and the idea that there exists an uneconomic growth that occurs "when increases in production come at an expense in resources and well-being that is worth more than the items made."

> Environmental degradation is an iatrogenic disease induced by physicians (progrowth advocates) who attempt to treat the sickness with unlimited wants by prescribing unlimited production. We do not cure a treatment-induced disease by increasing the treatment dosage. (Herman Daly)

To counter economic development, we must develop a concept of anthropological sufficiency. The notion of "sustainable development" is merely a means of making the notion of development sustainable, although it really should have already been sent to the idea-recycling plant. But anthropological sufficiency does not mean *self*-sufficiency ("sustainability"), given that life is difference, is relation with alterity, is openness to an outside in view of its perpetual interiorization, and interiorization is always unfinished (the outside maintains us, we are the outside, we differ from ourselves at every moment). What is in question is self-determination, the capacity to define for ourselves a good enough life, as Winnicott spoke of a "good enough mother." We do not need paradise, or a perfect mother; the "good life" is a good enough life. There is no better than enough.

Development is always deemed an anthropological necessity because it supposes an anthropology of necessity: the subjective infinitude of man (his insatiable desires) is in indissoluble contradiction with the objective finitude of the environment (the scarcity of resources). We are at the heart of the theological economy of the West; we are at the source of our economic theology of "development." However, this economic-theological concept of necessity is, in every sense, unnecessary, what we need is a concept of sufficiency and not of necessity. Against the theology of necessity should be put forth a pragmatic of sufficiency. Against the acceleration of growth, the acceleration of transfers of wealth, or the free circulation of differences; against the economicist theory of necessary development, let us devise a cosmo-pragmatics of sufficient action. Against the world of "everything is necessary, nothing is ever enough," favor a world where "very little is necessary, almost everything is enough." As Anne Ryan wrote, Enough is Plenty. Who knows, maybe if we learn to abide by this principle we will end up with a world to leave to our children, and their children. "Had we but world enough, and time."

I conclude on a pessimist-fantasist note, by saying that I have my doubts that we can escape the ecological crisis created by capitalism simply by means of the exercise of scientific reason and political will. I have therefore come to suppose that only a religious movement, a posthuman messianism, biocentric and geomorphic, can perhaps modify the conditions of our existence in a significant way. It is a matter, then, of shaking the religious foundations of Western culture, and perhaps even of human culture. Humans must mutate into another species to forestall their own extinction. Christianity was a radical innovation within the anthropological matrix; it redefined certain basic values of human society. Christ was an anthropological Messiah. Perhaps the imminent *parousia* will bring us a different Christ: a physical Christ, a thermodynamic Messiah. But of course, messiahs are not made to order. Okay then, let us say a religious antifundamentalism.

DEBATE

BRUNO LATOUR

Talking about your presentation, you said there is no alternative to the anthropomorphic, which I understand is very different from the anthropocentric. Can you explain in some practical example what it means? Because from the outside it looks perilously close to the cliché view of the Indians as those who live in harmony with their ecosystem, they do not want to grow too much, they are very careful even though, of course, the reason for such a behavior—as you beautifully described it—is completely different from any sort of ecological thought. It is not out of respect for the environment and nature, it is, on the contrary, because they generalize anthropomorphism. But, in terms of diplomacy, of care and caution what are the consequences?

EDUARDO VIVEIROS DE CASTRO

When I said that anthropomorphism is the only livable alternative from anthropocentrism, I said livable not conceivable. There are many other conceivable alternatives. From an existential and anthropological point of view, humans will always be central in the sense that we have no other choice because we can only see the world from the point of view where we are. Humanity will always stay a human centered perspective but then you have an alternative, either saying that this perspective actually circulates in the universe and every being can take this perspective or you consider yourself as the only legitimate occupant of that position. I think these are the only alternatives. For Amerindian anthropomorphism, humans are not privileged. The human perspective is not privileged from a general point of view but it certainly is privileged from a human point of view. The Indians do not think that seeing the world like a jaguar is as good as seeing the world like an Indian. They say "It is as good as seeing the world like an Indian if you were a jaguar but we are not jaguars." They are very clear about that. It is not about everything; it is not a fusion or a sort of paradise in which every being is the same as every other being. Actually, what makes the world dangerous is precisely because jaguars are human not because they are not human. There is this Indian saying that Irwin Goldman, a very good ethnographer of the North West Amazon: "The fierceness of the jaguar is of human origin because jaguars do not kill people normally, if a jaguar kills a man you can take my word this is not a jaguar. It is a man. This is the man inside of the jaguar." So what makes the world dangerous is the fact that it is saturated with humanity, not because it is devoided of humanity. They are no *pris d'effroi devant ces espaces infinis*. They do not think that solipsism is the major problem. The lack of communication has never been the problem of the Indians. It is an excess of communication that is the problem.

ANNE MARIE REIJNEN

I won't let you off the hook so easily. I would precisely like to come back to that part of your presentation. First of all, I would like to say that I found it extremely promising, and I really hope that I will be able to integrate some of your perspective, but I have to be sure that I understand where the common place may be, where we meet.

Hence my question: is there a contradiction when you wish for the mutation from the human species into some unknown "X," and when at the same time you agree that there is no alternative for human beings for being human? Then there is also, it seems, a contradiction with wishing for a *deus ex machina*, when the overall category that you would like to impose or to propose is the "good enough." If it is good enough, then should we not do some *bricolage* with what is there? Specifically, you say in the beginning there was nothing, just people. And would you then accept that in the end there will be nothing, just people, rather than to wish for a thermodynamic Messiah, which would be I think to let yourself led astray by fanaticism or escapism? This is a good enough world. It can be a good enough world.

EDUARDO VIVEIROS DE CASTRO

That is a tough question because "in the beginning there was nothing, just people" was not a pleasant situation at all. It was just an initial situation. I think if in the end we get only people again, this is going to be a disaster and this is actually a disaster that is predicted time and again in Amerindian mythology. So it is common having Amerindian mythology about successive earths that get destroyed in a new earth coming over it because earths get excessively populated or people get too wicked. Due to these successive floods, successive destruction and creations, in the beginning and in the end we are going to have only people. What was the other contradiction, Anne Marie?

ANNE MARIE REIJNEN

Between the good enough and the second coming, the mutation, the post human. Which is your form of transcendence?

EDUARDO VIVEIROS DE CASTRO

How to become another species? Maybe there is a contradiction, I am not sure. If everything is human, as in Amerindian cosmology, then the human is an entirely different thing, to begin with. So it is not simply a question of generalizing the human condition. Once you do that, the human condition gets much more complex, much more internally complex, because it is not only the case that jaguars are humans at the bottom, it is also that humans have a jaguar inside. And so humanity becomes a far more interesting proposition.

ANNE MARIE REIJNEN

I am reminded of Giordano Bruno who said that until we accept our animal nature, we cannot be human. Another illustration of the point I made earlier: there is so much in the Western tradition—the Heterodox thinkers of the Christian tradition—that is there and that we just need to bring back into the limelight.

EDUARDO VIVEIROS DE CASTRO

Yes. We have taught in our sociology classes to think that humans project their humanity unto the cosmos. What I think Indians do is exactly the opposite: bears and jaguars are as much human as humans are bear like, and jaguar like. They also introject the cosmological qualities in themselves.

BRUNO LATOUR

Do I understand you right that shamanism, this capacity of introjecting human qualities distributed everywhere, including inside humans, is really for you an alternative to nature? That it would be better than Creation or creativity to establish the basis for diplomacy? And that the reason is that it would oblige everyone to be extra careful, to exercise care, because it would make you be especially attentive to the encounter with all sorts of other entities?

EDUARDO VIVEIROS DE CASTRO

Absolutely. When I said that we have no alternative except anthropomorphism and anthropocentrism what I was actually hinting at is the idea that the only proper way of actually feeling the need to have care for creation is to infuse it with human qualities and not with human dangers. I insist Indians think the world is a very dangerous place to be, precisely because it is saturated with humanity, and our world nowadays is saturated with humanity in a slightly different sense, but in the sense that it is totally technologized. We live in a very similar world to the Indians, in a slightly different way, different not metaphorically but literally. I think this care should be lived as a kind of consequence of the anthropomorphic principle. The anthropocentric principle leads you inevitably to carelessness, to mistakes, *gaspillage*, to waste. I prefer the "good enough" life, a concept that I love. Winnicott's idea of a good mother is a good enough mother, a perfect mother is a terrible mother. A good enough mother is what a child needs to be properly transformed into a proper human being. Just a good enough mother. Do not over do it. There is no better than enough, and I think we should try to apply this principle to mother nature as well, to Gaia as well. A good enough Earth would be all right. A good enough life rather than a perfect life, better and better, every time better life that would lead us to a psychotic situation, a collective psychosis. It is precisely what happens when the child has a too good mother. I think our Western civilization is a perfect case of collective psychosis, due precisely to our striving for the perfect mother or father.

ELIZABETH THEOKRITOFF

I am totally fascinated both by your paper and your oral presentation. And as somebody was quoting yesterday, I could also say that you are telling us our story. I am constantly reminded of what I've read of the mythology and traditional world of the Alaskan peoples—not that I am expert on this, but I know something about it mainly from the work of Father Michael Oleksa, who is an Orthodox priest who has served most of his life in Alaska. His wife is a native Alaskan; as you probably know, the Alaskans are for the most part Orthodox Christians. But in a very interesting way, much of their traditional worldview has quite seamlessly been taken into their Orthodox Christianity: and I am extraordinarily struck by the parallels between the worldview, the understanding that you are describing, and that which is basic to Orthodox Christianity. Even the idea that "at the beginning there was nothing but people." What do we Christians say? "In the beginning there was the Word." The Word is the prototype, the archetype according to which man is created "in the image." The Word is also the one who will become man. The image, the reflection of the image in man that we talk about can also be described as "rationality." It is the same word in Greek, *logos*. When you say "rationality" in English today, people think of it in a very narrow sense. It is commonly taken to mean a capacity for speech and discursive reasoning that distinguishes man from the "dumb beasts." But an integral part of the traditional Christian understanding of man's rationality is that, because we are in the image of the divine *Logos*, we therefore appreciate the *logos* or rationality that is in everything because everything was created through the Word (*logos*) of God. In other words, we live in a world in which every thing images the Word in whose image we ourselves are made. Now, that is not the same as saying that everything is human; but there is, as it were, a triangle, in which all creatures are connected by the Word who is also the archetype of man. Our humanity and rationality refers to the Creator Word; and everything else also refers and relates to the Creator Word, rather than directly to us.

I also wanted to comment on the idea of a soul in all things: again, I remember Father Michael Oleksa writing about the traditional Alaskan depiction of the sun and moon with a face, and he comments that this is not anthropomorphism in a negative or crude sense. It is saying that in a certain sense, these are personal beings as we are. Obviously, in the Christian tradition we do not believe that the elements are literally animate. We do, however, believe that they too reflect the Word through whom they were created.

It almost seems that we also have in your presentation the idea of the "garments of skin," in the sense that I discuss in my paper, when we say that death is both a consequence of a mistake and also a blessing, because otherwise the Earth, as we know it, would become unlivable. And indeed, it seems to me that there is some parallel with the cosmic ramifications of the fall in saying, for instance, that the fierceness of the jaguar is of human origin. As I said, the parallels are really quite extraordinary: there is so much here that can be taken up, echoed, and understood on the basis of Christian cosmology.

EDUARDO VIVEIROS DE CASTRO

You make a distinction that I didn't make. I said everything is human. Actually, the proper translation for this very common statement in Indian languages is "everything is people," rather than human. Precisely because the notion of human species is not exactly what is at stake. Everything is people, everything is a subject and is capable of subjectivity. When a jaguar looks at the mirror or looks at another jaguar, he says "humans are anthropomorphic entities." So, people see themselves as humans, including us, we have no privilege. Humans see themselves only as humans but all the animals disagree with us, about us. Our essence is actually very much at risk: we do not know who we are, that depends on who is looking at us. We think we are human, jaguars think so too.

ERIC GEOFFROY

Is the idea of "Thermodynamic Messiah" coming from your own imagination or does it rest on some shared beliefs?

EDUARDO VIVEIROS DE CASTRO

I think this is as good a prophetic ending as anything else. It is not a joke at all or it is a serious joke. I do believe that we need something equivalent to the revolution that Christianity introduced in human history but no longer with a focus on moral interhuman relations that is typical of Christianity. It has tended more and more to be that, in the sense that Christianity has become more and more inwardly, so leaving the world, abandoning the world. Not Christ but Christians. What kind of God will save us? I would say only a thermodynamic Messiah, meaning a very physical one. One who is really worried about the second law of thermodynamics and I do believe that science will not get us out of it. Politics will not get us out of the mess if defined in too narrow terms. So we do need some sort of new messianic movement. Messiahs are not made to order, we can't tailor them, we can't engineer them; there are no genetically modified messiahs (GMO).

I do think that we will eventually face the explosion of a myriad of messianic movements all over the planet, all of them having a very strong ecological component. Just like Zimbabwe people, and this is probably going to burn like wild fire if things start getting either really or imaginarily worse. We are going to see lots of candidates to be the new thermodynamic Messiah and of course the Jews will say of them all: "No, this is not the true one."

ANDREA VICINI

You emphasized the role of the subject. From an anthropological point of view, my knowledge of ethnic groups, however, places the subject within the community. The subject is inseparable from the community. Our own Western understanding of the person as an individual is out of place in these communities. The person always belongs to a network of relationships. She cannot be understood without those connections. Even in the expression "people" I understand a specific "community," a web of relationships.

EDUARDO VIVEIROS DE CASTRO

That is a very good remark, very perceptive, because indeed all these people are species. Subject is a society, it is not the jaguar. It is the jaguars as a society, as a collective, as a community. We are talking about collective subjectivities. That is why I emphasized the idea of kinship, having children, living in kin and with kin. Because there are no isolated individuals in this world. You are always part of a community. So, the idea that everything is human, everything is person, everything is people also means that everything lives in a community. An isolated individual is not human, you can only be properly human if you live in a society. The difference is that they think there are many different societies on Earth; the environment is a society of societies. There is no difference between society and environment, as the subject versus the object. Actually, the object is just another subject or collection of subjects. There are subjects everywhere and all of them are collective. Indeed, it is a community based view; Indians are communitarians of a sort.

ANDREA VICINI

As a consequence, our salvation will not come from an isolated Messiah. But it will depend on how we become more and more *one* people, an interconnected community, even an interconnected web of species within creation. Otherwise, it seems that we are alimenting an unrealistic expectation that comes from an unrealistic understanding of reality. We should avoid expecting that only God will save us, that only human beings will save us, that only technology will save us, and so on. By risking to oversimplify the complexity of reality, the history of anthropology, the history of humankind, and the history of religions point us in the opposite direction: when we will be an authentic society, an authentic community, we will fully save us.

EDUARDO VIVEIROS DE CASTRO

To wait for an individual Messiah seems to me not only nonrealistic but not very efficacious either. That is why all of us are betting in the multiplication of messianic movements all over the world. We will start claiming messianic messages. We will always be preaching a radical reorganization of society as well. This is what messianic movements are all about, about the reorganization of the societies.

ANNE MARIE REIJNEN

I was thinking about the one and the many. It is a common misunderstanding about Christianity that we are a monotheism. We are not! The God of our faith is a Trinity; God is sociality. The relations between God and humankind through Jesus the Christ is expressed by Luther's *communicatio idiomatum*. It is also expressed in the orthodox idea of Trinity as the *perichoresis* that is persons of the Godhead going round, a holy dance as it were. You keep talking about Christianity as a monotheistic religion but we have a Trinity. I will let you sort this out with the Jews and the Muslims, in the sense that they seem to be very monotheistic. Still, I do not think three is enough as far as the many is concerned. I think three is very little. I would rather go for the many in an indefinite sense. What we need is an infinite number of persons, of divine persons. Three is already a way to show us the way, three is not a number. That is a very naive understanding of the Trinity. The Trinity is a structure.

EDUARDO VIVEIROS DE CASTRO

Are you saying that there is no monotheism at all? Nowhere? I find it very disturbing and at the same time very comforting.

MATTHEW ENGELKE

This is a fascinating discussion. I am learning Anne Marie is Durkheimian and that Eduardo is a Hindu. But I want to get back to the point about Christianity as a radical innovation within the anthropological matrix. Because this relates to the appeal by certain atheists or philosophers, among them Alain Badiou and Slavoj Žižek, to the Christian tradition, and particularly to Paul, precisely for this kind of claim to radical innovation. Paul is a model of radical innovator. I wonder what is the radical innovation of Christianity.

EDUARDO VIVEIROS DE CASTRO

I think there are better qualified people around this table to tell us about that but Christianity invented a new concept of humanity, and it is probably the only radical moral innovation in the history of the West. It is interesting that Christianity came from the margins of the Empire, and that the idea of thermodynamic Messiah came to me when I was reading the book *Empire* by Toni Negri and Michael Heart. And I was thinking who is going to be the Christian or the Christianity of this new modern empire, who is going to play the role that the Jews or the Jewish sect played in the Roman Empire, which was completely subverted from the inside and was co-opted by the

Empire. That is why it reached universal domination. But let me be frank on how to define in a few words what is the radical innovation of Christianity: a radical innovation in the anthropological matrix.

SIMON SCHAFFER

I think political economy is the most radical innovation in the Western tradition. And it dwarfs Christianity and all other religions. It systematically refused to make a move many apologists often make, which is to say: "do not look at what my believers do, look at what they believe." Political economy says exactly the opposite: "Do not look at what my people believe, look at what they do." That is the opening not just of Adam Smith's *The Wealth of Nations* but also of Adam Smith's most Amerindian work, *The Theory of Moral Sentiments*. This latter book was Adam Smith's attempt to work out how the exchange of perspectives makes societies and *The Wealth of Nations* works out how the exchange of goods makes value. Those are the two key aspects of political economy.

NOTES

1. This text was written as a reply to a number of statements made in 2008 by the then Extraordinary Minister for Strategic Affairs of Brazil, the political scientist R. Mangabeira Unger. During Mangabeira's brief ministerial mandate, Marina Silva, the Minister for the Environment, was forced to leave the government, thanks to the aggressive anti-environmentalist policy of the ruling faction of the Workers' Party, which counts in its ranks both President Lula and his Chief of Staff Dilma Russeff, now the leading candidate to the next presidential elections.
2. Actually, Unger said that Amazonia was "more than a collection of trees; it has people in it" —people who needed state-sponsored development brought to the region, of course. So Amazonia is, according to the minister, a collection of trees plus a number of human subjects of the State who would rather be anywhere else than amidst said vegetal collection.
3. Amazonian rainforest is a rhizomatic assemblage—let us recall that trees in the Amazon region have relatively shallow roots, supporting themselves mostly thanks to their intricately interlocked superficial radicular system as well as by their enormous buttresses, and feeding to a substantial extent off of their decaying matter: rather than growing in the soil, they grow their own soil.

12

Economic and Ecological Challenges

Is There a Role for Religion?

Ignazio Musu

It is always difficult for an economist to speak to a multidisciplinary audience about ecological problems. Usually among noneconomists the perception is that economists, by preaching economic growth and by praising globalization, not only are neglecting ecological problems but are supporting an ideology responsible for the ecological crises of our time.

I had no difficulties in accepting that there are serious problems in the relation between economic growth and economic globalization on the one hand, and the ecological problems and the ecological challenges of our time on the other. The historical experience is that particularly in the take-off phase of economic growth, and therefore in those developing countries that turned out to have been successful in their growth experience, there is a statistical relation according to which economic growth goes together with environmental degradation. There are practically no exceptions. Among the most recent and impressive examples, are the past 20 years in China. Although the second world economic power, China is still a developing country that had a tremendous growth performance and at the same time sent frightening signals of environmental disasters. Only recently, and with enormous difficulties the Chinese ruling class became aware of the need of a drastic change of direction.

Should we conclude then in favor of renouncing to economic growth to avoid environmental problems? Should we enter into a perspective of renouncing to economic globalization for the same reason? I would be very careful in arriving to a conclusion like this. We cannot deny that economic globalization was able to allow billions of people to leave absolute poverty, enter into development, increase their average level of life. In China, 400 million people sorted out from absolute poverty since the moment economic reforms started in 1978 after Mao's death.

Of course, globalization and growth increased relative inequalities. Moreover, we are all familiar with the financial unbalances of the recent economic crisis. But I personally would be very careful in concluding with a negative assessment of economic globalization. Also I notice that there are contradictions in the present debate on the relation between economic growth and globalization on the one side and environmental issues on the other side. Let me give two examples of these contradictions. First, we are witnessing that in Europe, carbon dioxide emissions are now slowing down. Are we happy about that? Probably not so happy if we realize this is due to the economic recession and the accompanying increasing unemployment. Second, we may be in principle against economic growth, but we ask, now, to countries like China, India, Brazil to grow because we, in our mature countries, are unable to sort out of the economic recession: we would like that these countries grow in order to allow unemployment in our developed countries to reduce.

But let me come to the most important point. In principle there are no reasons why there should be a contradiction between economic growth and environmental quality preservation. We know that when we have economic growth, the scale of the economic activity widens and the pressure on the environment increases. But in principle there are no reasons why this should not be compensated or even more than compensated by, for example, new technologies reducing these pressures or by changes in the structure of consumption reflecting preferences oriented toward a more environment friendly direction. In principle, there are no reasons why economic choices and policy choices affecting economic behavior should not reflect some kind of appropriate value given to environmental preservation, expressing a greater willingness to pay of the society and its members for protecting the environment. There are no reasons why in our economic choices we should not give value to the life of nonhuman beings, to the preservation of species, by declaring our willingness to pay for that, which simply means to be prepared to give up other choices that are against this result. This may be true in principle. But why does not this happen in practice? Economists have pointed out some factors why this does not happen. The most well known is that when we try to improve the environmental quality we are just trying to improve a "public good." When somebody contributes to improving environmental quality, this benefits everybody, even those who do not contribute. This is a serious incentive to "free ride" about provision of environmental quality. It is a perverse incentive to irresponsible behavior and this is the reason why the majority of economists have been proposing and suggesting public regulatory policies to deal with environmental problems, based on making people aware of the costs they are imposing to the society when they degrade environmental quality with their polluting behavior.

So, we need public policies; some of them may use specific economic instruments: price incentives, such as taxes on pollution or tradable emission permits. However, the experience also tells us that, although some countries turned out to have been more successful than others, in terms of improving the environment quality, environmental policies are not so encouraging. The question is again: why? In general I believe that the need of corrective public policies would be reduced and also environmental policies would be more effective if society's members were motivated in their preferences and choices by moral values so strong to countervail and to compensate the perverse incentive to free riding. This would give not only more role to the market in orienting economic choices toward environmental friendly solutions, but would make citizens more willing to support environmental regulation and to respond positively to it with their every day behavior. Economists when dealing with public policies tend to use some kind of cost-benefit analysis. But when we come to the issue of how to implement the procedures for this analysis, we immediately realize their weaknesses. Some of these weaknesses are precisely connected to the conflict dimension of ecological problems. Take the example of a developing country where a cost-benefit analysis comes to the conclusion that the economic benefits of industrialization implying the destruction of a forest area are very high. The question is: which type of costs was considered. How was the value of wiping out cultures of indigenous populations, animal and vegetable species, assessed? Was the analysis caring for this loss of natural and social values? Probably nor, particularly it was inspired by powerful people living far away from the potentially destroyable area. A comparison of benefits and costs turns out to be very often incomplete and strongly biased when it proceeds without listening to different positions, without any respect to indigenous traditions, often making violence to them.

The point is that when we face conflicts, we implicitly face different ways of considering values when addressing environmental problems. The only way to reduce conflicts is through moving toward more widespread shared values. But here we encounter another problem. We cannot be satisfied with mere declarations. It is not enough to declare and claim that we have values that dictate that we should protect the environment. We need behaviors consistent with that. This is the only really successful possibility to support policies and to change economic choices. We need correspondence between declarations and behaviors. This is clear when we look at the global ecological problems. In this field we even lack the possibility of a public international policy. We simply do not have environmental international authorities. We need to recur to agreements between sovereign nations. On the other hand every global ecological problem has a conflict intrinsic dimension. Take the example of climate change. You have developing countries that point out that the per capita emissions of advanced countries are very high; hence, developed countries should act first. Developed countries on the other hand point out that the aggregate emissions of developing countries are increasing. China in 2030 will alone emit more carbon dioxide than Europe, plus United States, plus Japan together. Hence, developing countries should act first. We have a clear conflict dimension: a typical situation "a war of attrition" where no agreement is likely to sort out. The problem is clearly global: it does not matter from where carbon dioxide emissions come; every nation should contribute; cooperation is necessary. But to declare that cooperation is necessary is not enough; we need concrete cooperative acts consistent with declarations. But the incentives are perverse: they do not favor cooperation; they favor inaction. No surprise that international agreements are so difficult in this area.

Take, on the other hand, the biodiversity preservation problem. There is here the issue of the rights of indigenous populations on the biological resources used as inputs for new biotechnologies and new drugs. Often these resources are stolen through biopiracy. Who protects the rights of the genetic resources owners? No international enforceable agreement has been produced so far. The Biodiversity Convention is a framework waiting to be filled with some practical proposals. In some cases, the existence of a technological solution may be helpful to solve the global problems. Why was Montreal Protocol so successful to deal with the ozone depletion problem due to excessive emission of CFC gases? Simply because a technical solution, an alternative to CFCs, has been found by Dupont Corporation. Given that the technological alternative existed, it was easy for countries to agree on adopting it. For the climate change problem things are much more difficult. We do not have alternatives to the use of fossil fuels; we must rely on further research. Probably cooperation should be directed more to technological cooperation to find out a technological reliable alternative. A big internationally coordinated effort for a radical technological "break through" along the lines suggested by Nordhaus and Shellenberger is the right path to be followed. Of course there will still be problems concerning the costs of the alternatives, who is implementing them, who is paying for, who is acting first. But the existence of an alternative is a sort of necessary, although not sufficient, condition for successfully addressing these other problems linked to a conflict relation among the concerned countries. Even if we believe that the way out is to cooperate for an alternative technology system able to replace the current energy system based upon fossil fuels, we need people convinced that the scientific effort to find out the new technologies is worthwhile. We need people around the world, we need ruling classes of nations, convinced that it is important to invest in research and technology in this direction rather than in others. We need an ethics that consider worthwhile spending time and resources in inventive activities to obtain the required new technology results favoring a long-run solution to ecological crises and conflicts. Shared ethical values are crucial. And behind shared ethical values we also find religions. This is where religion comes in. Can religions help? I do not know. Some religions are more prepared than others concerning ecological challenges. Take Catholic religion, for example. It is true that in the past few years, its more or less official documents are giving some attention to ecology, but sometimes one has the feeling that this is made to respond to the current demand of the public opinion in the churches; and I do not see much practical behavior in that direction.

In monotheistic religions the fact that nature is considered as "creation" is important. A believer feels more strongly committed if he understands that it is the willing of his creator that he cares about nature and creation in a constructive way. The Christian message adds to this the responsibility of believers in re-creation in the perspective of resurrection. This may be a strong incentive toward changing practical attitudes toward the ecological challenges. But global ecological challenges of our time are not limited to the areas where monotheistic religions are the most important. The most important ecological problems now involve East Asian countries, such as China and India. For some years I have been working with Chinese researchers and civil servants on the issue of sustainable development. China is an example of ecological disaster but it is also an example that times are changing. The new ruling class, the presidency of the "fourth generation," has proposed some typical slogans, based upon the ancient Chinese culture: one is the idea of an "harmonious society" that requires harmony first of all within the person, then with others persons, and finally with nature. Nature is a totality; and the word "creation" does not have a particular meaning in the Asian religions. In China the ruling class is trying to use the slogan of an harmonious society also in order to convince people that they should learn how to balance economic growth with some perspective of social justice and attention to ecological problems. Where does this idea of harmonious society derive? It derives from new Confucianism, a mixture of ideology and religion. Zhang Zai, during the Song Dynasty (960–1126), introduced the concept of *qi* (vital energy) so central together with other concepts such as *yin-yang*, for a cosmological Confucianism able to recognize the balance between humanity and nature. In a famous "Western Inscription" Zhang Zai writes: "Heaven is my father and Earth is my mother and even such a small creature as I find an intimate place in their midst. Therefore that which extends throughout the universe I regard as my body, and that directs the universe I consider as my nature. All people are my brothers and sisters, and all things are my companions."[1]

To conclude, to the perception of the globality of the ecological challenge, and to the need of common values as a basis for solving conflicts and implementing cooperative strategies, must correspond a search for a sincere dialogue between religions, not only the monotheistic religions but all religions. This dialogue should be encouraged because it will be fundamental in reorienting practical human behaviors toward a more consistent model, in encouraging human ingenuity toward finding the appropriate scientific and technical solutions, in transforming tensions in the international relations into opportunities of global cooperation.

DEBATE

EDUARDO VIVEIROS DE CASTRO

My question has to do with the notion of political economy as such. Because when I studied economics there was a radical difference between economics and political economy. It seems to me that Ignazio is speaking from an economics point of view, not from a political economy point of view, in the technical sense. I would like to know more about this difference because economic concepts—like opportunity/costs analysis—do not belong to the tradition of political economy. In the political economy language, there is a much stronger emphasis on social transformations, on the changing of the relations that constitute societies. So, I would like to know what are your sentiments, your feelings about that difference. Second question: what is the political economy doctrine concerning the thermodynamic component of the material world? Because it strikes me that economics is the less materialistic science I have ever studied: there is no reference to physics in economics, no reference to thermodynamic limits, no reference to quantities of solar energy as constituting constraints in any given material production system. They never entered the Earth and they never left. There was no sun coming in and no Earth. There was just input and output and the waste went nowhere. What is the thermodynamic component in the notion of economic growth?

IGNAZIO MUSU

The political economy approach is extremely important. I myself have tried to implicitly introduce it in my presentation when I claimed that a cost-benefit analysis although necessary is not sufficient because it is biased against aspects concerning precisely political economy issues, such as distributional effects, conflicts, racial effects, impacts on cultures. Actually, what I tried to do when insisting on the conflictual aspect is precisely introducing some elements of political economy in a purely technical cost-benefit analysis. I don't think that the political economy approach excludes categories such as, for example, opportunity-cost, or, to use the classical language of David Ricardo, of comparative advantages. The point is that these concepts are essential tools of analysis, but in a wider political economy approach they must be considered as useful instruments to assess alternatives, not in an exclusive way.

In focusing on the alternatives among which a choice must be made, and in finding out the appropriate alternatives to be chosen, we should take into account not only the purely economic aspects, but also the social and cultural ones. This has an impact on how the cost-benefit analysis is used: benefits and costs are not only the purely economic ones, those that are measured through market transactions; moreover, monetary valuations, although not merely based on market information, may prove to be ineffective. Something more complex is then required, such as dialogue, search for a common basis of ethical values; these type of problems are not always solved by recurring to the advice of experts.

This is particularly the case in developing countries, where peasants and other weak social groups, who had previously been successful in establishing a good relation with the environment, are negatively hit by choices of industrialization and urbanization. Clearly here we have a clash of cultures, and at least an attempt should be made to address this clash, avoiding that policy choices are made by the most powerful economic forces, often using violent methods.

It is clear that the more diversified is the cultural attitude toward environment in different groups of a society, the more difficult is to implement successful public policies in favor of the environment with a wide degree of support from the society. Factors able to drive the society toward a more homogenized cultural vision in favor of environmental protection are therefore very important. A widespread social environmental responsibility goes in this direction, although it should not be seen as a substitute to public policies; rather as complementary to them and a way to make them less invasive and more effective.

Coming to the other question, about the neglect that economic theory has shown toward the implication of taking into account thermodynamic principles, there is a big debate between a stream, which we call environmental economics, which is directly inherited by the neoclassical theory and, what is called ecological economics. Ecological economics developed precisely to fill the gap created by the traditional neoclassical environmental economics approach that ignored the implications of thermodynamic principles for the economic analysis of ecological issues. The role of ecological economics has been crucial and we now see how environmental economics, especially when considering intertemporal problems, such as the relation between economic growth and environmental preservation, or the issue of optimal use of renewable natural resources, does often consider the implications of thermodynamics in describing the relation between the economic system and the ecological system. The debate progressed, and there is now an attempt to build a bridge between the two disciplinary approaches.

The traditional "economic circle" considered in the initial pages of any macroeconomics textbook must be widened in the sense that instead of having closed circles between consumers and firms, in terms of goods and money, you have to widen them and to introduce environment as a source of inputs and as a receptor of waste; thermodynamic principles introduce two types of constraints in the working of circular relations. The first is that matter can only be transformed, not expanded; the second is that in the transformation processes you need energy, and energy dissipates producing increasing entropy.

This way of considering things has profound implications. For instance, we need to change our idea of growth when considering thermodynamic principles. Growth simply cannot be material expansion. It can be "value" expansion, given the amount of transformed matter. This value may contain an increasing share of immaterial production with a qualitative upgrading of the same matter produced and used. On the other hand this value should be appropriately reduced to take into account the value unrecycled wastes, lost environmental quality, and losses of entropy. From this point of view GDP is not an appropriate measure of growth.

A very important contribution from ecological economics has been the attention given to ecological conflicts. An economist of the *Universitat Autònoma de Barcelona*, Joan Martínez Alier, a representative of the ecological economics school, recently wrote a very important book *The environmentalism of the poor* (Edward Elgar) where he analyzes different case-studies of ecological conflicts, taken from Latin America and Africa, where indigenous population and especially the poorest are seriously damaged by choices of environmental transformation and destruction. Economics should take more seriously into account these types of analysis.

ELIZABETH THEOKRITOFF

The question of behaving with consistency, of putting beliefs into practice, is obviously crucial. So I want to talk about this, and some of the problems it presents specifically in connection with environmental problems. Let us focus on the very basic Christian principal of behavior, love of our neighbor. Of course, one can produce endless examples where this is not implemented; but it is a principle that one can consistently appeal to, and measure one's own behavior against. And, concretely, this has always had in the Christian community a very specific economic manifestation in terms of giving to the poor, collecting money for those in need. It seems to me that one of the real problems in getting Christian individuals and communities to act according to the principle of "love your neighbor" in response to environmental and economic needs—because I do not think that one can separate the two—is that the world today is so complicated. If you tell a congregation: there are people who are starving, who are homeless, who are in extreme poverty, will you give some money?—then they will. If you start trying to explain to them that if you make this or that economic or environmental decision, it is likely to benefit other people, then it is much less probable that any action will result. I think there are two reasons for this: (1) the connection is often too indirect; the chain of cause and effect is too complex and involved for people to grasp; and (2)—here we come back to the question of conflict—there will be legitimate disagreements over which action, which decision is going to be the most beneficial to other people. You choose to buy an organic product, say, so that you are supporting organic producers; but on the other hand, someone will say that the people working in the sweatshops also need a job, that a sweatshop job is better than having no job. And for that reason, I think that very many people who might want to act according to love of neighbor end up simply throwing up their hands. It is too complicated to decide how to act. Anything I decide is going to be bad for somebody: so I might as well just buy what I like. This, I think, is one of the real practical questions we have to address: How do we teach people to see the connections between the principle of "love your neighbor" and our everyday actions and decisions? It is not that people lack the motivation to take practical action consistent with their faith; but rather, there is a sort of paralysis resulting from not knowing which practical action actually best reflects the love of neighbor. I do not know whether you can offer any helpful guidelines on that.

IZABELA JURASZ

While listening to you, I have the feeling that the problem is located between the scientists of ecology and the economists. In this case, the debate shouldn't take place between ecology and religions but rather between the economists and a materialistic and utilitarian mentality. You have said that economy doesn't rely on spiritual and cultural values and this gives indeed an interesting insight on the values and answers that religions can deliver to the questions raised by the ecological crisis. Is the absence of this spiritual dimension an important absence in the relationships between ecology and economy?

IGNAZIO MUSU

The first question concerns the difficulty we meet when trying to translate into practical behavior sentiments that we derive from our religious conviction, such as, for a Christian, love for others. What does it mean to love others when we face ecological conflicts? How do we respond to the protest of those who are loosing their jobs because the firm in which they work is closed down for ecological reasons?

It is very difficult to answer in a clear-cut way, as it is always difficult when we face choices affecting the life of other persons and their families. When we have to choose we must introduce relative values; absolute values do not help. We must create priorities; and doing this always brings the risk of overvaluing something and undervaluing something else. Hence, I agree, it is very complicated.

Sometimes, however, it is possible to have "win-win" solutions. This happens when a more efficient allocation of resources may be possible; by "more efficient" I mean a different allocation that is able to benefit at least somebody without damaging nobody else. This may be the case of absorbing workers fired because firms are closed for environmental reasons in other environment friendly activities.

The probability of "win-win" solution is higher in the long run where all the advantages of a new more environment oriented kind of model of economic development can show up. The likelihood that this new model will be implemented is higher, the more the connected political choice is supported by a convinced majority in the civil society. In the short run however is it more likely that trade-offs emerge, such as the negative impact of environmental policies on employment; policies should give attention to how to deal with these short run trade-offs and to appropriately compensate the losers, whenever these compensations make sense, which is not always the case, especially when values not transferable into money are at stake.

The experience shows that dealing with the short run costs in order to obtain larger future benefits from a new model of development is easier during time in which the economic performance is thriving than in periods of economic recessions and crises. In periods of economic difficulties, such as this period of unsolved financial and economic crisis, people are reluctant to accept changes in the economic model; fear is prevailing, and selfish behavior dominates. People become more conservative. Everybody wants to preserve its own threatened wealth; solidarity is more difficult. The work of altruistic religions is also more difficult.

The other question concerns the excessive materialistic attitude of economics, its difficulty of accepting moral and spiritual values. In my opinion the problem here is the following: is this reluctance intrinsic to economics or is it characteristic of a particular historical economic realization? The question is very old: do we need to be selfish in order to promote exchange and division of labor that can produce good economic results for everybody? The need of specialization does not derive from selfishness; it simply derives from the fact that it is impossible for an individual to do everything. We are limited. Everybody should do the things that he is relatively better to do. But that does not mean that greed is the fundamental motivation of individual economic behavior. Historically it may well be the case that Adam Smith was right in claiming that it is more likely the human being helps the others by using his self-interest, than through benevolence: But it may also be the case that selfishness is not the prevailing motivation. And in any case the experience shows that the conditions for a selfishness based market exchange to produce common good outcomes are so strong that they never realize in practice. This does not mean that self-interest cannot be useful as a tool for developing a personality; but certainly self-interest is not an absolute objective, and should be controlled and corrected so as to become instrumental to helping the others. This corresponds to the increasingly recognized need of an ethic dimension in economics.

In the ecological issues the problem is made more complex: we should not only care about the negative effects of our behavior on the environment because this may damage others (as it happens when their health is damaged by pollution); but we should also be concerned with the negative effect of the ecosystem components such as animal, vegetables, and biodiversity. The horizon of the required benevolence and hence of the moral dimension is widened.

BRUNO LATOUR

My question tries to relate to *oeconomia* as *pronoia* plus a lot of other things in Patristics. "Economy" in the sense that it has been invented by Smith is not a bifurcated concept. I remind you that God plays an immense role in Smith since it is the connector between moral sentiment and commerce or industry. Is there any way in which *oeconomia* might be the alternative to "nature"? Because that is the question we are trying to find out. In this dialogue we try out so many other words: creation, creativity, cosmos, shamanism, perspectivism, moral economy. We are trying to find the words that could be used as the umbrella for the diplomatic work of the future. Is this *oeconomia*? In the twentieth century, economics was the *lingua franca*, is there any chance that *oeconomia* be that new common tongue?

IGNAZIO MUSU

This is very crucial point. Considering economic activity, and any other human activity, as separated by nature implies a big risk. That of considering nature, the "natural status" as something external to the human activity, something that must be respected as such, and hence preserved as such. This is the big problem with environmental fundamentalism. Human beings are part of the nature: they act on nature; they transform it. One of the favorite slogans of environmentalists in Venice is that we should respect the "natural status of the lagoon." But what does this mean? The Venice lagoon would not have existed without a decision of Venice inhabitants to preserve it. There is nothing "natural" in the Venice lagoon; it is a nature that results by the past actions of the human beings, and to be "preserved" will continuously require further appropriate action by future human beings.

But if human beings can act and modify the environment in which they live, this means that they have a responsibility; this is the basis for recognizing that economic activity, when dealing with ecological realities and problems, cannot avoid assuming a moral dimension. The methodological individualism and the neopositivistic approach of the dominating neoclassical school in economics is guilty of having ignored this requirement of morality that we find in the "classical" economists, in Smith's *The Theory of Moral Sentiments*, but also in some "ecological" passages of John Stuart Mill's *Principles of Political Economy*.

Categories such as "public goods" or "public bads" that are typically applicable to ecological problems and challenges, although they come from neoclassical economics contain an implicit moral dimension. Environmental quality as a public good means that only relying on selfish motivation will not work because each individual will try to benefit from the virtuous and costly behavior of the others without paying any cost. These categories are useful when deciding how to implement environmental policies. However, the issues at stake when dealing with ecological global crises is wider, a deeper awareness of the need of a moral dimension in economic behavior and in the organization of economic strategies (including institutional adjustment) is required. The moral dimension has to do with the way creativity and innovation is mobilized in order to promote the environment oriented type of technological progress necessary to reconcile economic growth with preservation, and possibly improvement, of the environmental quality.

BRUNO LATOUR

Let me push that. Because "nature," as Michael Polanyi showed so well, is not actually *res extensa*. It was invented by economists as that which politicians should not put their hands on. It is a defensive concept to protect from politics. My question is: do you feel comfortable with a moral economy? I think that the moral dimension is absolutely essential if we want to get out of this contradiction.

ANDREA VICINI

Often we risk reducing our reflection on sustainability to a moralizing approach. Such an approach does not address the complexity of the issues at stake. First, do economists discuss the real impact on the environment of our personal behavioral choices? A hermeneutic of suspicion would suggest that these choices are not the real solution.

Second, to avoid any moralizing tendency, we should distinguish among the types of discourse that we articulate. When we reflect on behaviors and personal choices, we should appeal to values that concern personal development, which point to the personal way of interacting within society. At the same time, we need good proposals in political economy, as well as technological proposals. We want each discipline to address specific and problematic areas in light of their competence. Finally, we want religions to be involved but not merely moralizing. Hence, my concern is: are we mixing the discourses? How can we try to see what each discipline and each areas of human research can offer to the others?

IGNAZIO MUSU

Green consumerism is important. If people change their preferences, and hence their consumption and lifestyle in an environment friendly direction, this will help the market itself to work for solving ecological problems. We often argue in the following way: what will be the real impact on environment of my virtuous behavior if everybody else behaves in an environment unfriendly way? Clearly if everybody argues in this way, nothing will never happen. Things will not change. I believe that small steps are important to create the atmosphere favorable to more fundamental required changes.

However, changing the structure of demand is one aspect of a strategy for a sustainable development. The other, perhaps even more important, aspect is promoting an environment oriented technological progress. But the motivation for undertaking the huge amount of research required to lead to such an outcome also comes, at least in part, from the awareness that a market for new technologies will exist, a market large enough to repay the big necessary investments. It is very difficult to convince private but even public initiative about the need and the importance of a strategic change toward a new technological framework without people recognizing that they have to change in their individual behavior and that it is worth for each of them to contribute to supporting, eventually with the taxes that they will be going to pay, a strategy of environment oriented investments for the whole nation.

Look at what is happening in the United States. President Obama was desperately trying to move toward a green economy but what is happening suggests that this will not remain a great expectation; it seems that it is going to turn into a big illusion. This depends very much on what types of policies people really want to support, including policies required to achieve the technological solution to the environmental challenges. The problem is to have "firms" and public investments to implement the required research and to embody it in new technologies. This will not happen if the society's members do not express clear signs in this direction with their behavior as consumers and with their voting behavior.

Moralizing the economic behavior in favor of the solution to ecological crises must therefore imply acting on values leading both to a more environment friendly consumers' behavior and lifestyle, and to the awareness that a global strategy is required to change the system of production, including things produced and technologies used to produce these things, in an environment oriented direction. The mobilizing role of religion could be important for both achievements.

ANDREA VICINI

During our dialogue, we indicated that the rituals are one of the possible ways in which religions can help mobilize social energies to promote sustainability. Within religions, these rituals are often forgotten. They are usually generated or demanded very spontaneously by the believers. We need rituals to mobilize religious passions in order to cope with ecological conflicts.

NOTE

1. J. Berthrong, "Motifs for a new Confucian ecological vision," in *Religion and Ecology*, edited by R. Gottlieb (Oxford: Oxford University Press, 2006), 251.

13

MODERNIZATION AS LIBERATION THEOLOGY

Michael Shellenberger and Ted Nordhaus

One of the most enduring characteristics of human civilization is the way ruling elites espouse beliefs radically at odds with their own behaviors. The ancient Greeks recited the cautionary tales of Prometheus and Icarus while using fire, dreaming of flight, and pursing technological frontiers. Early agriculturalists told the story of the fall from Eden as a cautionary tale against agriculture. European Christians espoused poverty and peace-making while accumulating wealth and waging war. And today the world's most technology-rich consumers voice anxiety that continuing technological progress and consumption will be our downfall.

This essay builds on the arguments made by Bruno Latour and others that we should be suspicious of philosophies (e.g., Platonism) and religions that reject this world ("nature") for other words. Dealing with ecological problems ("crises") instead requires a belief system that embraces an evolving, developing Creation—human and nonhuman. "Moralistic, spiritualist, psychological, and, I would argue, scientist definitions of religion have," Latour argues, "led theology, rituals, and prayers to turn away from the world."[1]

While Latour is concerned with the ways religion turns away from the nonhuman world, we are concerned with a mirrored problem, the ways in which eco-religion turns away from the human world. Those most active in dealing with ecological problems like global warming turn away from, and often against, the process of modernization, viewing it as destroying nature. This view is characterized by feelings of both hope and dread. Global warming, on the one hand, undermines nature. On the other hand, Greens say, it could result in a kind of apocalypse (from the Greek for "revelation"), which could lead to a new, more natural world. This view animates mainstream green texts from Al Gore's *An Inconvenient Truth* to Alan Weisman's best-selling *A World Without Us*, a scientifically informed fable about the return of the nonhuman world after humans are gone from the earth.

These narratives help comprise an eco-theology that has become a dominant worldview of cosmopolitan elites in the West. Over the past ten years the eco-apocalyptic view spread from a minority of intellectuals to the majority of elites in the developed world. Barack Obama, Angela Merkel, and Gordon Brown at the United Nations Copenhagen Climate Conference talks to billionaires such as Google investor John Doerr and Virgin's Richard Branson to elite university students, professors, and educated liberals in Europe and the United States increasingly agree that some kind of collective sacrifice is needed to avoid the end of the world and/or create a new ecological one. Either peacefully through ecological enlightenment or violently through ecological disasters, green cosmopolitans increasingly came to believe that humans must move to a new ecological way of being because Nature demands it.

But while eco-theology preaches limits to consumption and sacrifice, its adherents practice consumption at ever-greater levels. Indeed, the most visible and common expressions of faith in ecological salvation are new forms of consumptions. Green products and services—the Toyota Prius, the efficient washer/dryer, the LEED (Leadership in Energy and Environmental Design)–certified office building—are consciously identified by consumers as things they do to express their higher moral status. And yet, the belief that we must radically curtail our consumption in order to survive as a civilization is no impediment to elites paying for private university educations, frequent jet travels, and iPads.

Hypocrisy has rarely been a hindrance to religion and may, paradoxically, contribute to its power. In not practicing what he preaches the religious man affirms his superior status. He is communicating that he should be held at a different standard than the herd. That he communicates this through his behavior even as he says that all are equal before the gods is even more powerful. Paying for one's transgressions through indulgences, carbon offsets, or contributions to the temple are all ways of affirming superiority.

But green anticonsumption consumerism is more than a contradiction; it is an obstacle to achieving the kinds of changes required to save Creation, including biological diversity, nonhuman habitats, and resilience to climate change. Those actions include things like replacing fossil fuels with low-carbon alternatives, increasing agricultural yields to increase food production and reduce habitat destruction, and accelerating development so the poor are protected from the vagaries and violence of climate change.

TECHNOPHOBIA AS NIHILISM

For much of human and prehuman history we have both celebrated and been wary of our tools. But technology anxiety sharpened considerably among developed-world elites after the Second World War. Technology anxiety grew in part to the awareness that a nuclear war could result in a nuclear winter destroying the basis for life on earth. But the anxiety quickly spread to smaller and more local threats: pesticides like DDT or genetically engineered plants destroying whole ecosystems; medical and health technologies reducing mortality rates and causing overpopulation; the Y2K computer bug bringing human civilization to a state of collapse; nanotechnology and robotics turning the universe into gray goo; the particle collider creating a black hole to devour the earth. The greatest of these anxieties today is global warming, viewed not as an unintended consequence of energy production and land use but rather a direct consequence of human hubris.

Technology anxiety purifies technology, describing it as unnatural, profane, and inhuman, and attacks the idea that new technologies can solve the problems created by old technologies. The solution to pesticides is viewed as going back to an earlier form of agriculture, not inventing less harmful pesticides. The solution to peak oil is viewed as building a society that does not depend on using so much energy, rather than inventing alternatives to fossil fuels. And it is common for green leaders, including even clean tech entrepreneurs and venture capitalists, to warn that technology cannot solve global warming, because a radical change in lifestyle resulting from massively reduced consumption is required.

Narratives of technology anxiety hold many characteristics of a religious discourse. Technologies are treated, like Eve's apple of knowledge, as the cause of humankind's fall from grace. The problems from technologies are not unintended consequences of our ancestors trying to better provide for their children but rather punishments ("Gaia's Revenge") for having violated nature's laws. The problems caused by these technologies, and the development that results, cannot be solved by more technology and development. The result will be either catastrophe without redemption or an apocalypse leading to a new world.

These contradictions result in a kind of nihilism, defined here as the rejection of this world for a future world, whether heaven or *ecotopia*. The contradictions of this world will inevitably lead to a future without contradictions. The Bible begins with the book of Genesis hinting that something awful will come of those who eat from the fruit of knowledge and ends in apocalypse. In the run-up to United Nations climate talks in Copenhagen, Desmond Tutu, the head of the United Nations, business leaders, and others described global warming as a coming catastrophe that must bring us together as a single species. Greens led a campaign stressing imminent apocalypse, "Tck Tck Tck," linguistically close to "tsk tsk tsk." And the United Nations branded it climate talks in Copenhagen as "Hopenhagen."

Nihilistic religions are concerned with building institutional power, not solving the problems with which they claim to be concerned, perhaps because solving (rather than moralizing against) problems would diminish their power. The result has been that the problems to which religions speak are solved by secular not religious institutions. Family stability increases in wealthier societies and classes. Agricultural managers continuously improved pesticides, fertilizers, and seeds, resulting in more technological inputs and higher outputs. Overpopulation was not solved by environmental groups advocating forced sterilization and cutting off food aid but rather by agronomists such as Nobel Peace Prizewinner Norman Bourlaug, who saved more lives than any other human in history. Y2K was fixed by computer programmers. Stable states and markets have reduced poverty and to the extent that religion played a role it did so not by preaching apocalypse but rather thrift, hard work, and savings.

While technologists and technocrats deserve our gratitude for their discoveries and inventions, they have not offered a compelling new narrative much less theology to rival nihilistic eco-theology. On the contrary, much of the technology community, from inventors to investors to consumers, hold deeply nihilistic views. Others adhere to a kind of technological libertarianism, which denies the role of collective (state) investments in the basis for our prosperity, and thus constitutes a different kind of nihilism.

MODERNIZATION THEOLOGY

A new ecological worldview must reject the nihilism and technological anxiety that has characterized postwar environmentalism. It should begin with a sense of gratitude for this world, which was created by shared investments in innovation by our ancestors for the most humane of reasons: to deliver to us longer, healthier, freer, and more fulfilled lives. This modernization theology should abandon the concepts of sin and transgression. Ecological and social problems are unintended consequences of technology and development that can be solved or managed with more technology and more development. Technologies should be treated as human creations and continuously improved (Figure 13.1), not rejected as soon as their flaws are exposed.[2] Our technologies have always been natural, from animal skins and fire to solar panels and nuclear plants. And in harnessing discoveries wrapped in mysteries we should view our technologies as sacred, not profane.

This is modernization as a liberation theology. The liberation is never complete or pure but rather impure, tentative, and unstable. It suggests that various kinds of liberation—from hard agricultural labor, from high infant mortality rates, from tuberculosis, from oppressive family values—bring all kinds of new problems, from global warming and obesity to antibiotic resistance to alienation and depression. In attempting to solve these problems we will inevitably create new ones. These new problems will largely be better than the old ones, in the way that obesity is a better problem than hunger, and living in a hotter world is a better problem to have than living in one without electricity.

The benefits of modernization to nonhumans are not as obvious though still imaginable. Humans have domesticated ecosystems for our own use. It is habitat destruction not global warming, which is driving species extinction and the loss of biodiversity. The solution to this "crisis"—the domestication of earth—will come from kinds of modernization that may trouble adherents of eco-theology. Pulling agriculture back from forests will require increasing agricultural yields through genetic engineering. Replacing land-intensive

Figure 13.1   Armin Linke, École Polytechnique, *Alps model*, Lausanne (Switzerland) 2001, courtesy the Artist.

cattle ranching may require growing meat in laboratories, a prospect that oddly inspires more revulsion than does its contemporary death-intensive methods. Stabilizing emissions may require large numbers of nuclear power plants both to replace coal plants as well as to pull carbon dioxide out of the atmosphere. And the solution to the species extinction problem may be the creation of new, synthetic organisms no numerous or bizarre than those created by prehuman Creation.

Modernization theology's own ambitions are, at once, large and small. Modernization theology will not satisfy those seeking to feel powerful and important by moralizing against sins and predicting apocalypse, but it may stand in contrast to nihilistic theologies, depriving them of their power. Modernization theology tells a grand narrative of human evolution and development, but it is not a total narrative, explaining the solution to every problem in advance. But for all of its limitations, modernization theology may be just what the world needs as people in the developed and developing worlds seek a better story to provide their lives with meaning and guide the difficult choices we will need to make as shepherds of Creation.

TED NORDHAUS

I am going to start out addressing the nature of the ecological crises: We have spent a lot of time here talking about what religions and religious passions might have to offer us in addressing those crises, but we have not actually spent much time talking about what they really are. What is the nature of these crises? What is the nature of the crises that we might want religious passions to mobilize various human and nonhuman resources to address? And I think it is our view very strongly that the ways in which you understand those crises is extremely determinative of the kinds of conclusions you draw: both about what needs to be mobilized and how, and then what role religions and religious passions might play in that effort. I think one of the tensions that have bubbled throughout the dialogue is really a question of the ways in which the effort to deal with this is really a fundamentally creative enterprise. As Bruno has pointed out a couple of times, the religious passions that built extraordinary spaces and places and structures like Venice, imbued with technology and arts. To the degree in which this is an enterprise in moral instruction, the central challenge here and hence is what we might expect the religion and the religious passions to instruct. I would just take a couple of minutes to go through what we would call the basic math of the ecological crises. And I am going to focus a lot on climate. But I want to make clear that I could be talking about any of the great global ecological crises, whether it is habitat destruction or the loss of our global fisheries. The basic dynamics are very similar actually to what we see in climate. And climate is interesting because this has been quantified in some interesting ways that can shed some light on what the nature of the challenge is.

I am going to start with something called the Kaya identity, which is a very famous equation developed by a Japanese engineer, that really defines essentially where carbon dioxide emissions come from or what drives the level of carbon dioxide that human societies put into the atmosphere. The basic factors are global population, output per capita, carbon intensity, energy intensity. We can see here the combined trajectories of those four factors and the emissions that result from that. So you see output per capita rising, you see population rising dramatically, you see energy intensity decreasing. This is the measure of how energy-efficient our economies are. So, by 2035 we would have improved our energy efficiency almost in half, even if our global carbon dioxide emissions have increased by about 50 percent. And then you see the question of carbon intensity, which has been very slow to change over many years. And I showed this to make a couple of points. The first is that actually Ignazio pointed out that economic growth brings inherently with it—has at least brought inherently with it—greater ecological impacts. However, it is important to note that the output per capita has increased massively, both because we are wealthier and because our population continues to expand at enormous rates. So, when you think about what you have to do to address this issue you either have to do something about the trajectory of population (and consequently the trajectory of output per capita) or you have to drive profound technological change. I am going to put the question of population aside, if people want to come back to it they can (but we have had certainly very little success in recent years in actually having much impact on the trajectory of population). Then the question is: can you deal with this by dealing with output per capita or do you have to deal with this through technological change? So here is where we start some of our math. The issue put on the table by a couple of people here is that we really do not need to consume—at least in the developed world—as much as we consume. And instead of this worshipping of economic growth maybe we need degrowth. There has been widely cited research on at what point being wealthier does not make you any happier. Now this is very contested research, and in fact there are several very large global studies that have largely debunked what has been the excepted orthodoxy around this stuff. For those who have accepted it, an annual income of about $15,000 is what In eeded tob eha ppye nough.

For the purposes of this example I am going to accept it. Let us look at what happens to global GDP and global emissions if we just bring everybody in the world up to this basic level of what has been argued is reasonable sufficiency (Figures 13.2 and 13.3). You can see here population rising from 6.9 to 8.3 to 9.1 billion people over the next decades. Let us presume, for the sake of argument here, that income is reasonably and equitably distributed at $15,000 a year, which of course we know it is not now. And look at what happens, in order to just support in 2030, 8.3 billion people at a level of reasonable sufficiency: the global economy needs to more than double, and obviously that continues to go up as population rises in the continuing decades. Now let us take the International Energy Agency's assumptions about carbon intensity and energy intensity, and let us assume that we are still getting these remarkable results: actually, a 40 percent reduction in the energy intensity of our economy is an extraordinary reduction. So let us then apply that to our reasonable sufficiency and see what happens to emissions. We see here emissions in 2030 going up about 25–30 percent at a time where climate scientists are conservatively telling us that we need to reduce global emissions by 50 percent from 1990 levels. This just gives you a sense of the contrasts between a reasonable sufficiency economy for a population of 8.3 or 9 billion people and the emissions that would result from that. And you see just a profound disconnection there. I show this to really just to make

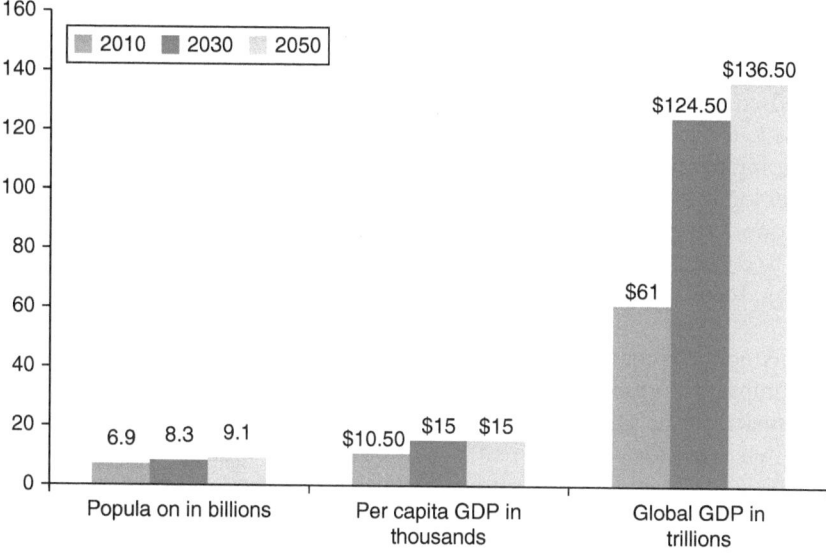

Figure 13.2 Global population and global GDP

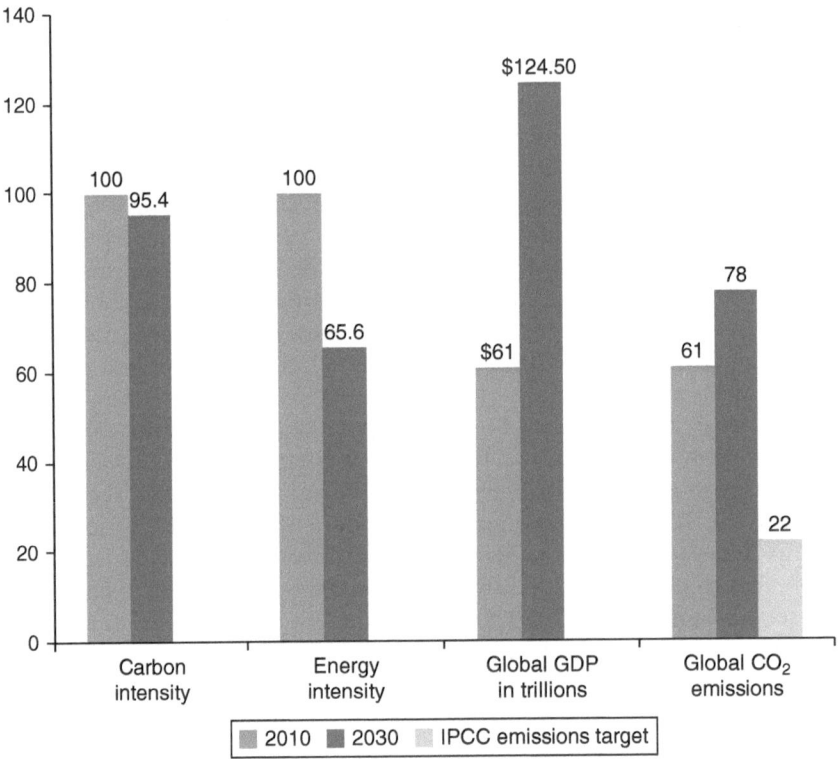

Figure 13.3   GlobalG DPa ndg lobale missions

one central point, which is that if we are going to deal with these problems it will be a fundamentally creative and technological enterprise. There is no other way to deal with this. Even if you limit everyone on the globe to this supposed level of reasonable sufficiency, you still see your emissions at three and a half times greater than what climate scientists tell us is the target. I am going to run quickly through the scale of the challenge, in terms of what technologically would be necessary.

Considering the global emissions in 2005, let us look at what it would take to cut that in half (which is actually less than what we are told we need to do). According to Roger Pielke Jr.:[3]

1. if Americans changed out all of the light bulbs—in order to get wonderful compact fluorescent light almost everywhere—that would have almost no impact on the global emissions;
2. if ten trees for every human would be planted additionally—we are all going to fill up our backyards and wherever else we can with trees—this would still have very little impact;
3. if all of these Americans who are driving around in their SUVs would be taken out of their SUVs, and put in the best hybrid electric fuel-efficient car we can, this would still have very little impact;

4. if we shut down every coal plant in India, then in the United States, and in China, we still would be less than halfway far from cutting in half the global emissions in 2005.

A presentation that the International Energy Agency gave at the Bali Conference showed what it would take to reduce global emissions by 50 percent by 2050: to build 30 new nuclear power plants around the world, then install 17,000 of the largest wind turbines you can conceivably build, now build 400 biomass power plants, two dams the size of China's Three Gorges Dam (a dam that is by an order of magnitude the largest dam in the world) and 42 coal or natural gas power plants with carbon capture and storage. Now do that every year, from 2013 to 2030.

So, that is what should be done if we take these threats seriously. Of course, this is a creative enterprise, but it also means that only technology will save us. And this has been our argument going back to our very first essay, which brought us to some prominence, which was the death of environmentalism. Contemporary political ecology simply cannot offer us answers to the new ecological challenges we are faced with. And no amount of reframing or inspiring people to change their personal behaviors in their everyday lives will ever cope with problems of this scale. It is a question of literally building an entire new society globally, just the very basic infrastructure of that society.

MICHAEL SHELLENBERGER

This is my first time in Venice and I have to admit that I was not prepared for my own reaction to this city, coming in from a rather hairy trip from the mountains into the city of water. Helen and I were in the back of the boat, just like children sticking our heads up and just really mouth agape at what a magical place this is. And the magic coexisting with a kind of decrepitude, the buildings simultaneously falling apart and being reconstructed. And when we went to Piazza San Marco for the first time yesterday, it was impressive to me that things were both shining and fading, the construction ongoing, while also the city was falling apart. And we think that it is a very nice metaphor for the responsibility that we have for Creation, both human and nonhuman. The problem that we address in our paper, given the scale of this challenge, is really a problem that Bruno has also identified: why have, especially the developed nations of the world and the elites in the developed nations of the world, been proposing a set of solutions that are completely inadequate to dealing with this problem?

And it is not just about changing the light bulbs, it is also a global treaty framework that would have virtually no impact on global emissions. Why has the emphasis been on reducing and limiting ourselves rather than creating something? And we hypothesize in the paper that this has more to do with who we are, that is "second moderns," following the sociologist Ulrich Beck. This has more to do with who we are as elites living in highly developed societies. Many of us, for example, do not feel the need to live in big houses and drive big cars; in fact we take pride in living in modest homes. We take pride in driving small cars. We take pride in eating less meat and taking the bus, and valuing a set of things that we imagine are not related to consumption. We say, "Well I'd rather be able to travel to Venice and experience the beauty of Venice, rather than buy a very big flat screen television set." But in fact, what we find is that everywhere the people who are espousing these solutions are in fact consuming, not just as much as they have always consumed but in many cases more, if you actually look at the amount of consumption that is going on. And we thought to ourselves: well, this is not just a problem of arithmetic; this is not just a problem of awareness; these are the facts that people have been saying for a long time. They have been laying out the problem.

So, we want to ask: what is it about this response to global warming that speaks to the age that we live in? And what we found was a whole set of values that we second moderns have and that we want to project out there as the solutions to the climate crisis. This leads to what we call a "nihilism," a negation of modern life, which is simultaneously amnesic and nostalgic. So we imagine that we are not living highly consumptive, highly technological lives. We imagine that we would be happier, and that we are happier when we do not live those lives. And we forget that the entire basis for these ideas and for ourselves is an intense amount of development of consumption over many centuries. We look at the Chinese and we say we want to help them not to make the mistake that we made in developing because look at all the awfulness that it has brought us. And we forget that really much of that development has been basic things, sewage systems, electrical systems, modest, basic levels of development that they are now pursuing. So why would we be in a moment where the wealthiest citizens of the world would be lecturing some of the poorest on how to live and how to develop? And what we want to do is to suggest that this is not an ironclad ideology, that it is unstable and that it is changeable. But that it is going to require that we tell ourselves a new story, both about where we have come from and where we are going. So, if there has been an older story that humans were created by a superior being, that we have fallen from an original unity and harmony, that we fell because we sinned against the laws of nature or God and that all of the problems that we experience from global warming to alienation are consequences of those violations and that as soon as we will align ourselves with those laws we will return to that original harmony and unity; then we want to replace that story with a new story that says: "in fact humans were created by earlier and lower beings who used technology to create us." They created weapons so that we could eat meat. They used fire to cook the meat, so we could absorb greater and greater amounts of protein. Our hips could get larger. We could develop more capacity for intelligence and control over our environments. They used animal skins so that when they gave birth to large-headed babies prematurely these babies could survive. They could wrap their babies around them with animal skins using technology in that way. And that we were constantly in the process of overcoming adversity, and in the process of overcoming those adversities we created new problems. And the process is never-ending. So we created global warming not because of cruelty or a lack of care or greed, but because we are trying to take care of ourselves and we are trying to take care of our children and our creations.

The solution to these problems is not going to be in withdrawing from that project of self-improvement and human creation, but continuing it. Therefore, we want to redefine the modern world, and here we borrow heavily from Bruno, as a world that remains enchanted and enchanting. The buildings of Venice are not profane, they are sacred because they were acts of creation. The fact that they are falling apart is not a sign of our sin, but just of the natures of this world that we live in. It would be, indeed, humorous for us to imagine that we could ever get out of this constant flux and cycle, that we could perfect our natures, that we could return to some original unity, that we are only gods in the sense of having extraordinary powers, but we are certainly not gods in the sense of being so powerful as to end this ongoing cycle of overcoming the new problems we ourselves created. We also borrow from Bruno this lovely idea of Doctor Frankenstein's sin, which was to abandon his creation rather than to care for his creation. And what might that monster have become had Doctor Frankenstein continued to care for him? And we see in this a metaphor for our own attitudes toward technologies that constantly disappoint us. Nuclear power plants have all sorts of problems, for example. But we are creating better and new ones and those nuclear power plants will both be our salvation and they will create a new set of problems. We really should appreciate the mystery in that, the contingency in that, rather than fleeing from it. And that this is what we are suggesting here is not a laissez-faire attitude. It is not an "anything goes" attitude. It is a deliberate, a thoughtful approach. It has a new idea, instead of thinking of sacrifice we want to think of investment. So, there are investments that need to be made to deal with these new problems, collective investments. We all need to pool our resources to create new creations just as Venice is pooling its resources to protect and re-create Venice. And we will need to do that with our entire world.

DEBATE

MATTHEW ENGELKE

I have one important question. It's important to recognize that one of the reasons why people invest in energy reduction, getting energy-efficient light bulbs or solar power panels on their roofs or hybrid Priuses, is because it makes them feel good about contributing to the world. But it's also at household level a microeconomic decision, which in the long run it's going to save this particular family money. So I switch off my television at the wall every night, not primarily because I want to save the world but because electricity in the United Kingdom is expensive. I'm not convinced there is wide spread evidence of hostility to development and technology. I certainly never heard

Bush, and I don't think I've heard Obama, say that we should get rid of our technologies. From everything that I've heard it's always about investing in technology. The dominant political narratives are not only that we need to use green light bulbs and fly around less, but we need to figure out ways of injecting carbon under the ground and coming up with new sources of energy. I don't see the abandonment of the technology that you herald. I don't see the political utility in the language of sacrifice, it never got anyone very far, except Churchill. And that was in a very unusual set of circumstances. Why has that narrative of sacrifice become so compelling?

TED NORDHAUS

Let me take the second question first. Because I think it's the easier one to answer and I might make Michael answer the first one. I think we would argue that you've gotten a sort of strange conflation of green limits focused politics and a kind of—at least in the United States, and I suspect in the United Kingdom as well—bipartisan neoliberalism, which we would also argue actually is in some part an outgrowth of second modernity. So, yes, Bush and Obama will never get up and talk about sacrifice. What Bush and Obama both do, even more so, much more so Obama, is that he gets up and says: we're going to have investment in all of these great new technologies and it's going to put everybody to work, it's going to make us safer in the world, we'll get off oil, and all of this. And then you look at what he's proposing to do and what he's talking about is private investment motivated by either regulations or carbon prices. This actually is a turning away from the role these technologies have always had, being largely invented and deployed by states. This was the result of collective investments and collective labors of society that was inspired, profoundly inspired, by the idea of building a new and better world. Today no one actually ever really deals with the real functional process of building the infrastructure of a new world. They're always being pushed down to the individual level: we'll put a price on carbon, whatever it is, and then all of you are going to turn your television set off and you are going to build solar panels. But this better world will not get built because we put a price on sin. I think that there are obviously many motivations for these acts. And I think what we get very concerned about, not so much in the hypocrisy of it, but in the ways in which it informs the larger political discourse, when you are told to turn off your television set to save the world and not to save money.

BRUNO LATOUR

This meeting was assembled around the question of scale, how can we scale up our reactions to meet the challenge coming to us, now and we have two scale points. One of them is the one you just showed here, the other one is the idea that the individual motivation and saving money and time were not enough. And that maybe there is something, some resource to be found in the scale of religious energy around the notion of Creation that could help us meet the challenge. But of course at the time when religion was much more pervasive in the life of people, for instance, when this marvelous place was built, it was not separated in religion, technology, science. My point and my question is really what is the difference between the technical fix argument and the technical care argument? In other words, you said technology is how humans do creation and salvation (Figure 13.1), which is a very powerful connection between the theory of technology and the theory of Christianity.

What I want to understand is what for you is the difference between the good and the bad technology. I think we all agree that technology is never just techniques but also a complex set of socio-technical assemblages. So what is the touchstone that allows us to choose between the two? If the hope that technology is the fix disappears, we cannot discriminate between good and bad techniques—that is, in the end between bad and good politics of development. If we don't have this touchstone we will enter into controversies about how many nuclear plants, which type of nuclear plant, and so on. We will be back to the situation where there is no solution, either in technology, nor in the market, nor in the invisible hands. And we will be as devoid of potency as before. Or, to put it another way, what is gained by bringing Victor Frankenstein back in his laboratory? If Victor Frankenstein stays in his laboratory what does he do that is different from what he did in the past when he abandoned his creatures? What it is to attend to unintended consequences? If there's a God in technology, which God is it and how do you follow his dictates?

SIMON SCHAFFER

What difference, if any, does "care" make to the socio-technical strategies we want?

MICHAEL SHELLENBERGER

I think that we tell ourselves the wrong story when our technologies fail. I think we look for some failure of intention about those technologies. And in the case of nuclear, the engineers were so excited about nuclear that in the United States in the 1950s they famously oversold the technology. But I think that they were not animated by an ethic that these technologies will fail, that there will be failure, and that the failure is inevitable, and that the failure itself is creative. The history of innovation is really a history of success upon failure upon success. So we discover things accidentally. We credit the critics of nuclear power for making nuclear power better. We imagine that some technologies will be sustainable, that they will solve problems and not create any new problem. We expect these technologies to provide us with things called sustainability rather than embracing the contingency of those technologies.

BRUNO LATOUR

I will follow that because that is one of the few topics on which I have some expertise. For 25 years I have been trying to teach engineers not to oversell the technology. Hype is part of engineering training, writing a PowerPoint, writing a business plan, doing bullets, and so on, is part of what it is. In other words, convincing technologists not to hype and oversell is difficult. What do we do about changing engineering culture?

TED NORDHAUS

I think that your friendly amendment to our argument about technology is that we reject the idea that these past problems are a result of a lack of care in a generalized sense. They are the result of a different care! What's going to make the engineers of today or the engineers whom we want to train different from the engineers of the 1950s? I would suggest that we care about different things now. And we particularly start to care about those things once a whole set of our initial material needs get met. Which is why even in the 1950s we were building cleaner conventional power plants and cleaning up water supplies and all sorts of other things. And that was all being done by engineers and often rather well. You can go to the San Francisco Bay, the Great Lakes, or any body of water in American now, virtually, and it's vastly cleaner than it was 40 or 50 years ago. And that is both a testament to the environmental movement, to critics of what the polluting technologies were doing, and a testament to the engineers who figured out how to put something on the end of a smoke stack or a pipe that prevented the pollution from going into the atmosphere or into the water.

SIMON SCHAFFER

You don't think it's mainly a testimony to the exports of large coal burning technologies from the Great Lakes region to the Three Gorges region?

TED NORDHAUS

I think it is not mainly a result of that. I would argue that when we look at particularly carbon dioxide emissions, the industrialization has had a huge impact. I would also mention the various claims made in the European Union that deindustrialization has been vastly more responsible for the supposed progress on carbon emissions than any of the policies that are held up.

EDUARDO VIVEIROS DE CASTRO

My first question has to do with the average income ($15,000) mentioned in your projection and your comment that obviously it's not equally distributed. I would like to know how higher or lower is it than the American average income of today.

MICHAEL SHELLENBERGER

I think that number in the United States today is around $40,000.

EDUARDO VIVEIROS DE CASTRO

To achieve that worldwide average income, you'd have to accept to reduce the US average income.

MICHAEL SHELLENBERGER

Well, let me clarify a couple of things. Let's put aside the fact that today global incomes are massively differentially distributed. Let's just look at what would the ecological footprint look like if we redistributed it all equally at $15,000 per capita to every person living on the globe. Let's also assume that if we're taking every American down from 40,000 to 15,000 we're really raising Somali up to $15,000 and we still end up with emissions that are four or five times above what they need to be.

EDUARDO VIVEIROS DE CASTRO

So, if we still will be above four or five that what we need, then we'll necessarily make the Americans go down four or five times lower than they already are.

MICHAEL SHELLENBERGER

Yes and no, and this is the second part of the argument, which is: when you actually look at the technological challenge associated with getting to where we need to get, even at $15,000, you must create new technologies that are consistent with everybody having an American level of income of $15,000 a year. I'm highly skeptical that, lacking some kind of very authoritarian effort, that will ever happen.

EDUARDO VIVEIROS DE CASTRO

Well, that's a kind of leap of faith. So, moving from $40,000 to $15,000 amounts to the same. The other point is that we couldn't reach this without authoritarian means, and I do agree with you. That was something I was wondering about. Granting that these are the measures that have to be taken, what seems to be missing from my point of view is the political technology that is necessary to implement this wish list. You said the American administration would be willing to do that but they want to make it a private enterprise. That is unlikely to happen because there are too many short-term interests. And then I was wondering whether you would not end up close to Lovelock's position, which is that we need tyranny in order to get Gaia back in shape. It would be a strongly state-sponsored effort that goes entirely against the spirit of the brave American private enterprise.

MICHAEL SHELLENBERGER

Just one bit of history in energy modernization. I'm not going to suggest it's the law, but many societies go from wood and dung as their primary source of energy to coal, to natural gas, to nuclear and renewables. Energy systems get modernized, and in that process governments are willingly paying a bit more money for cleaner sources of energy. That dynamic is happening; it's both through public policy and through the dynamics of modernization.

Now there are two countries in the world that have gone to low-carbon clean energy very quickly and as quickly as we need to go globally, and that's France and Sweden. And they did it through direct state deployment of hydroelectric dams and nuclear power plants. And they did so for non-environmental reasons, they did so for geo-political reasons. France doesn't have indigenous coal reserves. They didn't want to be subject to Russian gas tyrannies. And they have a lot of nuclear power plants and they have the most successful industry in the world. On the question of America, an interesting story is that Ted and I started our journey on this in the early 2000s, knowing almost nothing about energy or energy policy, but having been working for environmental groups who hired us to go and do public opinion research on how you get the public to support this. We went to the blue collar, rust belt cities of Akron, Ohio, and Pennsylvania and we started describing an industrial policy, which is a bad word for neoliberal elites. We started investing in factories and R&D institutes to build clean energy. And it was wildly popular. What was not wildly popular was making electricity cost more money so that people would be forced to use less of it and drive less.

America has always made huge investments in infrastructure and technologies beginning with Lincoln, the railroads, the highways. The Internet was invented, and there were also positive unintended consequences of Cold War investments. And those are the kind of investments that remain popular among most of the country except among the neoclassical, neoliberal elites who want to do all this by pricing carbon. And we think that, again, where you are seeing these big state investments in clean energy are in what Beck calls "first modern" countries, like China. China is going to be investing 70 billions a year in clean energy whereas we invest about 5–10 billion a year.

So the question for us as second moderns in the second modern society is how do we mobilize the religious passions to support this kind of "first modern" project in a "second modern" society. And just without going into a lot of details we tried pushing that first modern project and ran up against our green colleagues and our neoliberal colleagues, and hence our calling for the death of environmentalism.

ANNE MARIE REIJNEN

My question is one that's very brief. In two words: *cui bono*? *Cui bono*: that has been made very clear by each of the participants in these dialogues. We are coming from a discipline for which we have worked. We belong to a confession or another, or we study a field. I think if you have an argument, for example, against violent resistance, that argument is going to be very different if it comes from a pacifist than if it comes from a military lobby. The charts and numbers you have given us just now have this seeming clarity and objectivity. In fact, we all know that each of the numbers and their interpretations are subject to debate. So I don't have a problem at all with having advocates, or militants; I just would like to have you tell us clearly *cui bono*. What is your advocacy? What is your militancy? Make it clear who is funding Breakthrough. What are the bills that you would like to have passed in the current administration?

MICHAEL SHELLENBERGER

Who funds you? What's your interest? It's not clear to me.

ANNE MARIE REIJNEN

I work in two different places, both confessional. I'm a scholar.

MICHAEL SHELLENBERGER

Just so. We come out of the environmental movement, our parents took us to antinuclear rallies. Ted and I made our entire professional careers working for environmental groups. Our think-tank is funded by liberals and environmentalists. We are men of the left. And frankly, it doesn't matter.

ANNE MARIE REIJNEN

No man of the left would ever say that it doesn't matter.

MICHAEL SHELLENBERGER

Well maybe I'm trying to change, I'm trying to challenge the idea of what a man of the left is. There is a long tradition of the left that is a modernizing project, it actually includes Marx. Marx got turned into an antimodern sometime around the 1960s mostly by second moderns. Our project is a modernizing project. I think our intentions are about as clear as they can be. We tried to present our theology, tell you who our gods are, the kind of redemption that we would like to seek. It's hard to take it, a little bit like maybe you sense that ExxonMobil is behind our institute.

ANNE MARIE REIJNEN

No, it's not quite as naïve. And I don't think that you fully answered the question. Which is not a personal question in terms of who you personally are, I'm just trying to say that no one speaks really from Sirius, from an Archimedean point of view.

MICHAEL SHELLENBERGER

Right. And we would never suggest that we are. The fact that we use some quantitative analysis does not indicate that we don't understand subjectivity.

ANDREA VICINI

What is your political diagnosis concerning the coming years in terms of the US political response? What are we going to do about it?

SIMON SCHAFFER

Well, clearly it would be a great idea to remind Sarah Palin that the indigenous population of the state in which she used to be governor has an enormous amount in common with the inhabitants of the Amazonian.

ANDREA VICINI

First, during our *Dialogue*, we reflected on what we learned from the history of religions in terms of their concrete impact on cultures, society, and policy. We could also ask: What did I learn from the history of technology? When I am using technology and when I am thinking about technology in the future, with my human and intellectual skills, I would like to have learned from the past, by distinguishing what was wrong and what was right.

Second, we discussed Michel Foucault's biopolitics. Michel Foucault reads the history of science and of technology by focusing on controlling the body. Personally, I would extend this understanding of biopolitics to controlling everything that relates to people in the largest possible sense: human beings, animals, and also social dynamics.

Hence, how do we discern among current and future technologies? When we think about the future that we want for humankind, should we use a quantitative method? Should we use a method focused on efficiency? If we look at medicine, the criteria that we use to discern among technologies concern healing: Do they help people to be healed? Do they promote health care? What are the criteria that we could use when to reflect on the whole ecosystem? Should we use criteria based on avoiding harm?

MICHAEL SHELLENBERGER

On the first issue, when we speak of how second moderns get produced in this particular way, we pull from Foucault. Foucault is interested in the production of humans, not the control of humans, because he didn't want to suggest that there were these humans and then there were these external factors. As for the question of evaluating technologies, and in particular energy technologies, we look at energy technologies that have a modest, minimal impact on the land. So, one of the reasons why we have come back to nuclear as an important technology is in part because we get concerned about dams. We get concerned about how difficult it is to build very large solar farms, very large wind farms. And so this trajectory of energy modernization starts taking you to increasingly dense kinds of energy. There's a reason why in Antarctica the scientist stations run on nuclear power and not these technologies you have to spread out all over the place. We spent about the past six years advocating for a big wind farm to get built on the coast of Massachusetts, which was viciously opposed by the Kennedy family who didn't want any windmills in their viewscape. But it's taken them ten years just to get the permit to build it. It will take another five years to build it. And if you wanted to move the whole world to wind you'd have to build 14,000 of those wind farms every year. So, we start to look at things like nuclear again because it can be deployed quickly, at a low cost, with minimal impacts. And it's interesting the story of the Sierra Club in the United States, one of our most powerful environmental groups. It was pronuclear in the 1950s because they opposed the dams. And that only starts to reverse itself in the 1960s—actually before Three Mile Island and before Chernobyl—when a whole set of ideas about what was natural and what was unnatural, such as breaking apart atoms, came to define green thought.

I think there's a very common discourse that says that our technologies are racing ahead faster than our ability to control them. Another version of it is that our technologies have gotten out of our control and we lack the wisdom to control them. We don't believe that. Our view is that we've had the opposite problem: which is that in saying no to technological innovation we have made the problem worse. That's not to say that we shouldn't ever say no to any technologies, because we should. But it is to say that we have not had the ethic of caring for our technological creations as second moderns.

TED NORDHAUS

If I could just add a little bit more complexity to that, I think that we simultaneously have greater control over our technologies than we've ever had, and we have more control at the same time that we have more unintended consequences. It's both: the unintended consequences of that technology, that we extensively have more control over, are multiplying rapidly. And I think that in all sorts of ways, from the individual level to the societal level, we are putting constraints. The reason why we addressed so much to what we called an inchoate secular eco-theology is that certainly, in second modern societies people are really finding their spiritual identities and meanings much more in secular eco-theology than in any of these religious traditions, which in the developed world are all dying, whether it's Catholicism or mainstream Protestantism. And to the degree in which those religions are revitalized they're actually revitalized in more evangelical way, at least in the Christian context. One of the real challenges that we haven't really got to in this dialogue is that we are talking about how to harness religious passions and we're mostly having that conversation in the context of traditional theological frameworks of religions that are becoming increasingly irrelevant to the populations of developed-world societies.

ELIZABETH THEOKRITOFF

You say that reductions in carbon emissions are not going to be achieved by people changing their light bulbs. They are not going to be achieved by shutting down all the coal-fired power stations. So where is all the energy use? Where are all these carbon emissions coming from? And the second question has to do with new technologies. It is not clear to me: have you in mind technologies that are already known and that could be put into place? Technologies that one could start putting in to place tomorrow to cover present and future energy needs? And, further, do you envisage this process of new technologies to deal with new problems going on indefinitely? I know you're not fond of the idea of limits, but do you envisage that there could be limits on new technologies? I am thinking not least of potential limits on the resources needed to build new types of energy installations.

MICHAEL SHELLENBERGER

In fact, we do think that a large amount of emissions reductions will come from replacing coal plants. So you actually start to see big cuts in emissions from replacing the existing coal plants with clean energy.

ELIZABETH THEOKRITOFF

Relatively large, yes. But I am wondering where the majority of the energy use comes from?

MICHAEL SHELLENBERGER

Fossil fuels. So, just to oversimplify, 75 percent of our emissions are from the energy sector, 25 percent are from changes to land use, such as deforestation, but not just deforestation. If you look at the coal plants alone, it's huge.

TED NORDHAUS

And if I can clarify one thing, these numbers show shutting down existing coal plants. A lot of what you still have to do are the projected new coal plants that are going to get built. Because we actually don't have particularly scalable technologies that are real alternatives, particularly in the places where those coal plants are going to get built, which is China, India, and elsewhere in the developing world.

MICHAEL SHELLENBERGER

As for the second question (how does that transformation take place?), we should keep in mind that many countries move away from dirtier energy sources to cleaner energy sources mostly as a way to deal with air pollution. So China right now is investing heavily. They unfortunately don't have natural gas, but they're investing heavily in nuclear and renewables, mostly to deal with urban air pollution problems that cause many tens of thousands of deaths a year right now. The problem is that coal is very cheap. Energy is the mastery source. Everything in the economy requires energy. Coal is 2–3 cents a kilowatt hour, wind is 2–3 times this (plus, you have to pay for the batteries to back it up), solar is even more, 5–10 times the cost of coal. And nuclear is a bit more too. So, do we have technologies right now to replace coal? Yes. Do we have technologies right now to replace coal without creating a great expense and slowing the economic growth? The answer is no. Our country is willing to pay a little bit more money for energy, yes, not a lot more money: so, mass manufacturing small nuclear reactors rather than building expensive ones, mass manufacturing cheap solar panels rather than getting very expensive processes. So the goal is radical and is accelerating radical innovation, and the good news is that we actually know how to do that. We did that with microchips, we do with pharmaceuticals, we do it will all sorts of technologies. Our view—I mean generally the view among independent energy experts—is that in the United States we could probably use existing air pollution standards to shut down a lot of the coal plants. We're not running out of coal, we have somewhere around 200 years of coal left. And then there's a lot of disputes obviously about oil, but even if we do run out.

ELIZABETH THEOKRITOFF

I'm talking about the non-carbon fuels.

MICHAEL SHELLENBERGER

Oh, are we going to run out of silicon for solar panels or Uranium for nuclear plants? Is that the question?

ELIZABETH THEOKRITOFF

Yes—or whatever other resources may be needed for new technologies. But it is not just a question of whether we shall literally run out of them; it is also a question of whether the environmental cost of getting hold of them will escalate to an unacceptable level.

MICHAEL SHELLENBERGER

First of all, we will run out of these resources. The question is how quickly will we run out of them. There's really plenty of nuclear fuels, not just Uranium but also nuclear waste. There are now plants in development to burn old nuclear waste to power new plants. In fact, I think half of our nuclear plants in the United States are being run off of the old nuclear bombs. And there's obviously a lot of silicon and sand in the world. I think what we don't want to do is to just seem that any of this is easy or sustainable.

MATTHEW ENGELKE

One of the other questions I had earlier has been asked, but not directly. Anne Marie asked about your politics, but I wonder about the ethics embedded in your narrative of liberation—your modernization as liberation theology. You clearly have a conception of rights, understandings of freedom, self-autonomy, justice. So, classic modernization theory has a model, not only of technological development but also of political development. Which forms of politics are superior to others? This has been carried through into the neoliberal era, through the justification of the invasion of Iraq and also the war in Afghanistan. These were about giving people freedom—about Afghani women going to college and being able to listen to music. So that set of wars was fuelled by a moral discourse, an appeal to a moral discourse. I wonder if there's an equivalent here. What are the moral issues here and could the Breakthrough Institute imagine justifying military intervention?

MICHAEL SHELLENBERGER

We were both against the Iraq War.

MATTHEW ENGELKE

My point is drawing a parallel: could you ever see justification for war? What kind of interventions are you willing to make, not for freedom (because you don't want to talk about freedom) but only for birthrates and air conditioners?

SIMON SCHAFFER

Modernization goes along with a certain form of liberation theology; according to you, is there such a thing for the Breakthrough Institute as a "just" war?

MICHAEL SHELLENBERGER

I hate these "just war" conversations. I'm not going to have it, but what I would argue is that there are a set of preconditions for liberation that ought to be encouraged. Cheap, scalable, abundant, clean energy for the almost 2 billion people on the planet who don't have it today is a transformative technology that will drive some kind of modernization. What exactly that modernization looks like can be quite specific to the place. When I look at the way that the developed West imposes a set of modernizing discourses around things like human rights on developing societies, I think there's a bunch of ways in which it's quite problematic. I look at Africa and I see constant demands for more democracy and just an assumption that more democracy in Africa is the solution to failing states. And I have a hard time believing that this is the answer. I think that there are a set of preconditions for the functioning of democratic states that have to be in place first: basic security, a set of basic economic conditions, things like that. And I would reconsider more realistically a whole set of questions around human rights and development. Here we sit in societies that in most cases went through many centuries of brutal dictatorship before you could have anything that looked like political democracy, in the way in which it's been defined in the West. And it seems problematic that we then expect that somehow in Rwanda or Somalia or any number of other places, these Western style democracies are going to emerge with a little bit of food aid or some human rights sanctions.

MATTHEW ENGELKE

So, if I can just clarify: you would say that the development of good technology can happen at a quicker pace than the development of a good politics?

MICHAEL SHELLENBERGER

I would just say that those things emerge in very complicated relationships to one another. I do think that better technological solutions can make political solutions more possible in many cases.

BRUNO LATOUR

We had planned this meeting as an experiment, imagining that it would be interesting to bring the people from very different types of disciplines to tackle the question of scale in the types of emotions and passions necessary to handle ecological concerns. Now, after three days we might have an answer to this little experiment, which would be worth sharing with one another. One way to state it is to try to answer Eduardo's question: what's the name you give to the thermodynamic messiah? We might have completely different names, but we have a less difficult question if we ask ourselves: what is the name that could be used as the umbrella for creating the common world? Is it the word "technology," except it's a very different technology than in the past because it has got in it lots of politics and lots of techniques. Is it economy, moral economy as what was discussed by Simon? Is it Creation? Is it cosmos? If it is none of these, would we be considering that the experiment has failed? That we should not bring religious passion to bear on any of the serious ecological questions that has to be solved?

SIMON SCHAFFER

I'll say two quick things. I've been struck by an extraordinary ambiguity exactly on the question of scale since one is supposed simultaneously to think, as it seems, that the ecological crisis is unprecedented and absolutely precedented. And we're supposed to think both of those things at the same time: it is absolutely unprecedented according to the lines of the global maths that Breakthrough presented, a series of demands and challenges not much like what's been experienced in socio-technical development; on the other hand, it's absolutely like socio-technical experience in the past. Nothing new is happening here, not even scale. In fact, in a certain sense especially not scale, where scale is intensity as opposed to extension. That's to say that the quality of the crisis is more remarkable than its extension. The other thing that strikes me is that one is still not clear enough on whether the thought is that the ecological crisis is on such a scale that it requires the mobilization of religious passions or of a theology of modernization; or whether rather (which is in a way much more interesting) the ecological crisis is of a scale that it provides an absolutely splendid opportunity for religious passions, technological planning, social democratic, democratic centralists, state-driven investments, and so on. Those are not at all the same thoughts.

ANNE MARIE REIJNEN

Is there a consensus around this table that we are in this Dialogue talking about three, not two things? It is always helpful to have a ternary rather than a binary construction. Would you agree that we have been talking about beliefs? Practices? And that the third element is the common ground, which is the planet Earth. Is that agreeable to all? If I understand you correctly you also have a very strong belief in progress. So we have practices and beliefs and we all live on the planet Earth: that could be the umbrella for the dialogue. Beliefs, among which is the belief in progress, but we have other alternatives or things that are older than that. Anyway, I'm going to get back to the discussion: would you agree that there is a ternary construction here? The common ground being the planet Earth and then beliefs and practices that are going to shape our future on the planet Earth.

My second point is about what is unprecedented or not. I'm learning about this and I'm really looking for information, but is it not true that the global heating of the planet could lead to a tipping point that might be irreversible? And I concede it might not be entirely unprecedented, but it could be a mass extinction on such a scale that we will not survive to compare it with previous mass extinctions. That is my query.

EDUARDO VIVEIROS DE CASTRO

We haven't discussed the sense of urgency that we see a bit everywhere. Urgency in the sense that many scientists say that it has already happened, we won't change much, whatever we do. We can always obviously do lots of things, but there is something that we can't change, whatever we do. I'm not sure whether this is a correct assessment of the situation but it does connect with Anne Marie's worry, the widespread feeling of having reached a tipping point. We have already passed a lot of tipping points, and all the mass extinctions are already at our doors, and this is something that I haven't seen addressed here: we have talked about history, but past history, we haven't engaged in any futurology. What is going to happen? What's certain to happen? And what are we going to do about it, in all senses of the term? The horns announced the apocalypse. We are not hearing the fourth signs of a future apocalypse, the apocalypse has already happened. I would have liked to see this discussed. I'm not saying that it's too late for everything, I just want to know if it's too late for something because this has its theological implications as well. That was my only point.

PASQUALE GAGLIARDI

I would like to share with you some impressions "from outside," in a sense, because—though sitting around this table with all of you—my main role has been the role of the observer. If I compare this dialogue with all the previous ones, I have much more the sense that we are in the middle of the way. It seems to me that this dialogue has just started. Let's look once more to the key question we started from: "The gamut of passions mobilized by ecology so far is not at the level or at the intensity of what is required by the immensity of the tasks. Only religion seems to have mobilized in the past the right set of transformative passions. But it is not clear whether this level of energy is still available nor it is clear whether it is really useful. So, the key question may be to mobilize the notions, the rituals, the cosmologies allowed by some religious traditions, but on the condition that they allow the politics of ecological conflicts to be clarified." Clearly, the interdisciplinary debate helped us to highlight the complexity of the "chemistry" involved by coping with the ecological crisis, but my feeling is that the above mentioned questions remain unanswered. We could have scheduled a final session to find out some at least tentative conclusions, but there is no time left. I would say that the experiment is not finished. More than ever I think that this dialogue should continue, probably in a different way, taking some distance from the experience and trying to look at it retrospectively. Given the unprecedented nature of the crisis, we need an unprecedented method to cope with it.

NOTES

1. Bruno Latour, "Will non-humans be saved?," *Journal of the Royal Anthropological Institute* 15 (2009): 459–475, esp. 463.
2. Bruno Latour, "It's the development, stupid! Or, how to modernize modernization," in *Love Your Monsters. Postenvironmentalism and the Anthropocene*, edited by Ted Nordhaus and Michael Shellenberger (Oakland, CA: Breakthrough Institute, 2011), 17–25.
3. Roger Pielke, Jr., based on data from Maximilian Auffhammer, Richard T. Carson, "Forecasting the path of China's CO_2 emissions using province level information," Department of Agricultural & Resource Economics, UCB CUDARE Working Papers, paper 971 (Berkeley, CA: University of California, 2007).

Part II

Afterthoughts

A Preliminary Notice

Pasquale Gagliardi

A few months after the Dialogue, Pasquale Gagliardi started imagining the possible shape of a book drawn from it. Referring to his final remark in the last session of the meeting, he sent to the participants the transcriptions of their interventions, asking them not only to check and amend the texts but also to contribute to a sort of "virtual final session" aimed at taking stock of the whole experience. Therefore, he wrote to them the following letter: "Please, imagine that we are having the final session that didn't take place. I'm chairing the session and asking each of you to simply go back to the original questions—Can we mobilize the transformative energies of religions in order to cope with the ecological crisis? How the notions, the rituals, the cosmologies allowed by some religious traditions can be mobilized to this end?—and try to give 'your own' answer, whatever it may be. More generally, in this 'ideal' final session I would ask each of you to tell the others how this dialogue affected your thinking, what you learned from it, what new methodological perspective aimed at coping with the ecological crisis can be inspired—if any—by this experience. In other words, what is your personal outcome of this experience?"

In the following are the "afterthoughts" received. The reader will also find here a paper written after the Dialogue by George Theokritoff, who had attended the Dialogue mainly as an observer. The editors thought that including in the afterthoughts the reflections of a paleontologist would help the reader to set the present ecological crisis into a broader time/space horizon.

14

AFTERTHOUGHT: OF MAKER'S
KNOWLEDGE

Simon Schaffer

We have been invited to offer afterthoughts on the 2010 San Giorgio Dialogue: *Protecting Nature or Saving Creation?* The Dialogue's manifesto urged that to move on from somewhat sterile past exchanges between science and religion we should explore the tension between doctrines of nature and of creation. The high tension had somehow to be discharged because the sciences have apparently been able to demonstrate imminent ecological threat but not mobilize an adequate response. Meanwhile, religions have an impressive track record in mobilization, even if their capacity to get to grips with natural phenomena seems a bit uneven. The very first thing I learnt during the Dialogue was the intricate relation between the (somewhat religious) language of apocalypse and messianism, and the (somewhat technological) language of geo-engineering and planetary boundaries. As the *Economist* recently cautioned in its commentary on the arrival of the anthropocene epoch: "the invocation of poorly defined tipping points is a well worn rhetorical trick, stirring the fears of people unperturbed by current, relatively modest, changes."

I soon learnt, too, that the tension that affects relations between nature and creation relies in part on the notion of maker's knowledge. This is the principle that we know best—or indeed we only know well—what we make. One application of the principle is that creation's often been taken to be nature's making, so there is a long-term puzzle about the very status of nature as artifact. It is also a principle with some pleasingly posh and well-timed philosophical ancestry, much in evidence at just the period several historians judge the conflicts of western European natural science and religion began to reach toxic levels. It appeared in Francis Bacon's alchemically ambitious claim that anyone who can make a substance with all and only the properties of gold thus knows the causes of all gold's essential properties. It was pithily set out in Thomas Hobbes' mathematical and political polemic that the only "demonstrable arts" are those where the artist has power to construct the subject: "Geometry is demonstrable; for the Lines and Figures from which we reason are drawn and described by ourselves; and Civil Philosophy is demonstrable, because we make the Commonwealth ourselves." Hobbes reckoned natural philosophy undemonstrable because we did not make nature. And in his brilliant attempt at recovery of the oldest Italian wisdom, the Neapolitan friar Giambattista Vico declared "the true is what is made": because mathematics and civil society are human creations, they can be the subject of sciences.

Perhaps the speed with which I sensed how fundamental was the question of our best knowledge's dependence on the work of making was because my stay on San Giorgio started with a tour of the remarkable exhibition, *The Arts of Piranesi*, set up next door to the monastery in an exhibition space converted from the island's former warehouses. On show was a vast array of Piranesi prints from the Cini Foundation collections, alongside a multi-dimensional mobile digital projection of the delirious spaces of the *Carceri* and a set of material recreations of furniture designs taken from *Vasi, candelabri, cippi*, and *Diverse Maniere d'adornare i cammini*. The key (certainly ambiguous) word here is re-creation. The great Venetian architect was much affected by Giambattista Vico's contemporary arguments about the work of history and creation. In his own version of the inquiry into the oldest Italian wisdom, Piranesi famously declared it was never enough to imitate ancient models: the maker must display the possession "of an inventive, and, I had almost said, of a creating Genius." Neither the experience of sitting inside the penally claustrophobic flight simulator nor, even more obviously, gazing awestruck at stereo-lithographed realizations of Piranesi's ornamental designs, whether Etruscan vases or English coffee houses, were in any way recapitulations of events that had once happened back in the eighteenth century nor of objects that had ever quite existed in Piranesi's Roman workshops. On the contrary. These were in some sense unknown and unprecedented novelties, generated creatively through what its makers describe as a mix of organic modeling software and traditional modeling skills.

Many observers (such as, notoriously, the early nineteenth-century English opium eater and journalist Thomas de Quincey) had responded to the *Carceri* with narcotic reveries of flight through those visionary topographies. Yet none had quite sought to make them until now. Some of the images in *Diverse Maniere d'adornare i cammini* record finished work, but most viewers have instead been impressed by the ambiguities, inconsistencies, and downright confusions of Piranesi's shadowy project. Some have even judged the designs simply impossible to reproduce. Not so. Evidently the modern artifacts on display at San Giorgio cunningly conveyed forms of knowledge directly linked to the work both of past and of present artists. The nature of Piranesi's arts was marvelously grasped through these creations. Here was a rather dramatic and seductive demonstration that the capacity to make something was both a precondition of truly knowing the thing and also a graphic sign it could be known: thus a good start for reflections on the creation of nature and the nature of creation.

The theme of maker's knowledge mattered most in the Dialogue because of the consistency with which natural theology was treated as embarrassment, or failure, or demon to be exorcised. Some participants judged it a shameful episode in the science-religion story, preferably now to be surpassed. This natural theology seemed double-edged. On the one hand, it involved the claim that the order of nature taught humans how best to live; on the other, it insisted that the order of nature was a deliberately artful construct. Since my own contribution had involved a limited genealogy of the telling political claims both that nature should not be blamed for lethal environmental catastrophe and also that nature does not care for humanity, there were several occasions when it seemed important to explore both these senses of natural theological reasoning. The Patriarch of Venice reminded us right at the start of the Dialogue that it had long been recognized that the derivation of morality from nature was a mistake. So it looked as if humans could not learn how best to care merely by contemplating nature. Izabela Jurasz reminded us, too, that a distinction should be made between a maker who constructed a splendid artifact but then abandoned it to its own devices, and a maker who remained within the artifact engaged in the permanently careful task of maintenance.

Once again, these were arguments with very respectable ancestries, whether in Patristic theology or in notoriously tortuous Enlightenment debates on rival models of the permanent work of creation. Poets, it seemed, were artists who stayed within their work and sustained its life; clockmakers were artisans who deserted their watch once completed. This was exactly the matter of concern in early modern European controversies around a theodicy that drew much of its rhetorical power from shoppers' grumbles about faulty goods. "He who buys a watch," Gottfried Leibniz famously expostulated against the Newtonians, "does not mind whether the workman made every part of it himself, provided the watch goes right." According to the Hanoverian metaphysician, "creation wants to be continually influenced by its creator, but I maintain it to be a watch that goes without wanting to be mended by Him."

So I learnt that the all too familiar figure of the absent clockmaker still retained considerable vitality in discussions of ways of world making and therefore, by implication, on ways of caring for the world. This was something of a surprise. One of the more striking results of much work in sociology of science and technology is the insistence on the need for continuous repair work: instead of assuming local order possesses inertia, so that only major transformation requires explanation, it has been shown that the mundane and often unremarked work of fixing, adapting, repairing, and upkeep is persistently indispensable in maintaining anything functioning. When a system is maintained so well and so unremarkably that it seems to be entirely self-sustaining, it is easy to treat it as pristine nature. The hydraulic society of Venice and its lagoon is a compelling reminder of this fact: the price of artificial tranquility is constant vigilance. Our Dialogue spent much time

debating the effective technologies at work in sustaining systems that decep-
tively seem remote from the labors of engineering or economy. Technology,
we were instructed, is how humanity performs creation (Figure 13.1). The
cause and the response to ecological crisis reside exactly in this kind of art.
There is no doubt a politics here. The condescension with which repair is
treated discriminates against husbandry, housework, and everyday labor, and
discriminates in favor of moderns' heroics and showy *grands projets*. Perhaps
this was why the Dialogue closed with a collective voyage to MOSE, the
Modulo Sperimentale Elettromeccanico designed to safeguard the city from
floods by sealing the lagoon. But I missed that trip.

Thus a debate on creation, salvation, and protection quickly turned into
a conversation about the artificial character of nature. My written contribution had focused mainly on what I still take to be a key moment in moderns' development of these favored economic models of nature's artificial
capacities and threats: the Faustian conjuncture of global warfare, industrial
revolution, and capitalist transformation at the start of the nineteenth century. I was of course unsurprised to learn that this was also the moment that
some geologists have now decided marks the origin of the anthropocene
epoch, when humanity at last became a geological agent. I had also learnt
that the distinction between artist and artisan was supposed to map that
between vital creation and mechanical construction. It is no coincidence,
I guessed, that (in English, at least) this distinction emerged definitively
at just the same conjuncture. Before the nineteenth century, art was typically contrasted with nature, a distinction between the product of human
skill and the result of some inherent quality. Furthermore, the term artist embraced the sciences, technique, and industry: in London the mid-
eighteenth-century Royal Society for the Encouragement of Arts was a bul-
wark of industrial enlightenment. Piranesi was thus judged an artist in all
these senses: designer, producer, creator, and workman. But from the end
of the eighteenth century, elite usage began to credit artists with creativity
and imagination but held that artisans lacked these qualities; and from the
nineteenth century emerged the portentous distinction between so-called
creative arts and the realms of science and of industry.

There were obvious political-economic meanings in the development of
this telling contrast between high-flown spiritual creation and mere proletar-
ian labor. In reflecting on those meanings, I had referred to William Paley's
momentous formulation of the principles of natural theology in 1802. As is
well known, Paley's book starts with the image of a traveler coming across a
watch abandoned by the wayside, and the inferences about its artful maker
that could legitimately be drawn, even in the artifact's shamefully neglected
state, by examining its design. Natural theology warranted the inference
from art's divine cunning to the existence and the virtues of the artist. As a
kind of dark response, Mary Shelley soon composed her story of Frankenstein
(1818), a tale that eloquently charted the lethal effects on a creator were the
creation subsequently abandoned. Shelley recalled her own vision of a "pale
student of unhallowed arts" whose creation would "mock the stupendous

mechanism of the Creator of the world." Not at all coincidentally, she saw her own novel as a monstrous creation for which she nevertheless retained affection: "I wish my hideous progeny go forth and prosper."

According to the critic Franco Moretti, Victor Frankenstein's creature "is a pregnant metaphor of the process of capitalist production," in which the experimenter's art figures the emergence of an indispensable but hostile proletariat through the work of exploitation. At the start of the Dialogue, Bruno Latour told us that the fundamental question to be faced was precisely to work out the implications of a refusal to share in Frankenstein's abandonment of the results of this art. During the discussion of this theme, we reflected on the relation between stories of fatal abandonment and those of fatal control. We put the case of Prospero and Caliban alongside that of Frankenstein and his creature. Just as Frankenstein's monster's been taken as a figure of the proletariat, so Caliban's been taken as a figure of colonized peoples, condemned to subordination because of resistance to imperious power. At the end of *The Tempest* Prospero publicly recognizes the rebellious Caliban as his own ("This thing of darkness / acknowledge mine"), then orders him to menial labor. In contrast the virtuous spirit Ariel, who obeyed the rule, is freed. So alongside an interest in maintenance of care, the Dialogue began to address the need for liberation.

This was why it seemed like a good idea to reflect on encounters across starkly different cosmological frontiers, to understand how the enterprises of exploitation and economy work their effects. My brief stories of the fraught conflicts around observatories in Arizona and in Hawai'i were designed to address these questions. It might seem that talk of the spiritual values of astral science was simply a pragmatic response to Polynesian or Amerindian challenges to astronomers' work, a superficial and inconsequential move in a cynical political game. I was not so sure: rather, it seemed to me that these claims revealed much of what was genuinely at stake in modern sciences. This is why I was so impressed to learn from Eduardo Viveiros de Castro that the Kaxinawa people of Acre state in western Amazonia say "in the beginning there was nothing, just people." Speciation took place as the human aspect of the full range of beings moved into the background and they acquired bodies. And I learnt about these people's history. In the 1950s the Kaxinawa were forced into slavery to work for rubber producers, suffering systematic genocide, so that barely five hundred people were left alive. Much more recently, under the aegis of a nonprofit charity, named Living Jungle, the Kaxinawa have cataloged hundreds of potentially profitable species of plants and produce organic leather for expensive accessories marketed in Europe and North America by Hermès. Eduardo Viveiros de Castro also explained that Amerindians hold to a certain version of Malthusianism, preferring stable populations to the trials and strife of increased productivity and contestation. It was extraordinarily provocative to consider how these exploitative economies worked in a world originally made of human materials from which the range of species then emerged.

I was therefore troubled by a tendency in the Dialogue that generalized too easily and rather oversimplified what moderns' system of knowledge making involves. Consider one traditional account of the way the natural sciences make reliable knowledge of the world. According to that account, reliable knowledge is made by paring away whatever's unwontedly contributed by the observer, leaving whatever's real as a residue. This is why scientists often use a telling name for anything in their data they do not trust and treat as parasitic: they call them artifacts. According to this story, facts are made by unmasking and expelling artifacts. I recalled that those who have interests in denying the facts of anthropogenic climate change insist the data are artifactual: they spend a deal of time showing that climate models are manufactured. The strange thought, presumably, is that evidence of labor is always evidence of falsehood. And that thought relies explicitly on the dualist notion of a fundamental separation between humanity's passions and nature's condition. A nature that cares nothing for us would then be, precisely, a nature that could be known.

Yet, at the same time, the maker's knowledge argument urges that it is only artifacts that can truly be known. So for nature to be known it must at the very least resemble systems humans make. One of the best examples of the joint power of autonomous and of manufactured nature was of course provided by Darwinism. Charles Darwin's initial commitment to evident perfect adaptation, thus the need to treat its perfection as a puzzle to be explained, emerged significantly from the common natural theological and political-economic context of his epoch. He read both Paley and Malthus in the 1830s, returning to a reading of *Natural Theology* a decade later. He learnt from these priests that a system of benevolent laws generated perfect adaptation in all species. The organic world, he then cheerfully agreed, was structured as if it were an artifact. The task was to explain how that artful appearance had ever been accomplished.

It was one of the most intriguing consequences of these views that in retrospect it looked as if natural theology had been extinguished by evolutionary theory, when in fact it was one of its prompts. No doubt the ferocious public statements of a pair of pious and learned New Yorkers, Andrew Dickson White in 1869 and John William Draper in 1874, helped establish the conflict thesis at the basis of the story of science and religion. This is why it seemed important to recall that this story of a sempiternal conflict was an event within the history of modernity, not a moment of transition from premodern to modern worlds. I learnt distressingly little during the Dialogue of how religious passions might ever mobilize a considered and judicious response to ecological crisis or could ever overcome what Bruno Latour identified as the challenge of incommensurability: the evident gap between the scale of the crisis and the size of citizens' response.

This is exactly how Prospero is left at the end of *The Tempest*, deprived of all his slaves, reliant merely on devotion: "now my charms are all o'erthrown / and what strength I have's mine own, / which is most faint…Now I want / spirits to enforce, art to enchant, / and my ending is despair / Unless I be relieved by prayer." But I learnt a great deal that was fascinating and moving about the terms in which apocalypse emerged from religious reflection. There was also a lot to ponder in the implications of treating the world as an artifact. In particular, it became evident that it was remarkably disabling to imagine a nature or an economy entirely independent of and unresponsive to human purposes. What are needed are great ideas, just as Piranesi reportedly declared: "If I were commissioned to design a new universe, I would be mad enough to undertake it." In the epoch of crisis capitalism, it seems bizarre and vicious to treat the economy as a token of a natural order whose laws are fatalistically omnipotent and irresistible. And in the anthropocene epoch, with human and geological agencies so closely intertwined, it seems even harder than ever to disentangle a world that is made from a world that is known.

15

Thoughts after the Dialogue . . .

Izabela Jurasz

The four days spent at the *Fondazione Giorgio Cini* were particularly intense, whereas I am now thinking about them several months later. Of course that is quite normal. The number of issues raised and the high quality of the papers meant that it would be very difficult to reach major conclusions at the end of the Dialogue, even with the preliminary work. So in this sense there was firstly a very important mobilization of human energies.

Can we mobilize the transformative energies of religions in order to cope with the ecological crisis?

In a way this is a rhetorical question. Religions not only put forward theologies. They have always sought to answer questions concerning man and the world. Each religion gives different explanations about the relationship between this world and a divine being. Therefore in one way or another, they are capable of taking part in an ecological debate. But we can also say that the transformative energies are not only inherent in religions but have also always been inherent in the creation. I believe the real question is not about the possibility of mobilization, but the role that the religions already play in the ecological crisis.

The participants at the discussions showed that the impact varies greatly and that religious arguments can emerge on all sides. This reveals that the transformative energies from religions have already been mobilized. I would formulate the question differently. Instead of "Can we mobilize?" I would have asked "How can we use the energies that have always been mobilized today?"

How can the notions, the rituals, the cosmologies allowed by some religious traditions be mobilized to this end?

Here we have a real question...But it remains unanswered. I believe that we must start from the conviction that there are no more or less "ecological" religions. Each religious tradition can and must find in itself the resources—the ideas and cosmologies—that allow it to address the ecological crisis. Given the diversity of religious systems, however, there can be no single solution. The question must be asked of each specific religion.

The "religious authorities" would seem to be best suited to act as leading players in the ecological debate. Moreover, the participants at the Dialogue have highlighted that—at least in the case of the theologians—there has been an awareness of the ecological crisis for a long time. But this realization leads to another: the ecological debate is not a central concern for religion. Apart from the cult of Gaia, religious messages are not ecological manifestoes and it is unrealistic to expect a radical transformation in this direction.

We can ask ourselves then, is it really necessary to turn to the religious authorities to mobilize religious resources? Rather than mobilize the religions themselves, we must address the society in which they exist. In fact, believers are also citizens and religion is part of the cultural heritage of each country—even nonreligious countries. I think that in the environmental policy of each country one should take care to integrate religious elements and watch over the consistency between political discourse and religious message. This will obviously vary from one country to another. Yet without explicitly appealing to religious traditions the political authorities must take them into account. Arguably, the best mobilization of religious energies can be achieved in ways that are not at all religious. It would be interesting to explore the possibilities of integrating the wealth of religious traditions into the ecological debate conducted by the secular bodies responsible for ecological policy.

What is my personal outcome from this Dialogue?

Antiquity's leap toward the future—because it was a question of the future of the Earth—was a significant challenge. The first outcome concerns my own discipline. Rereading the ancient texts from a new point of view and communicating them to an audience unfamiliar with the subject was an exciting experience. Thanks to the dialogue with experts from very different fields we were able to raise some crucial questions, rethink methodology, and revise conclusions. It was truly very inspiring.

The second outcome is a little paradoxical. One might simply ask why introduce ancient texts into a debate on ecology and religion. Johannes Kepler was afflicted by myopia and diplopia at birth. His description of the heliocentric system with the planets in elliptical trajectories was based on reading Plato, before he found support in Tycho Brahe's observations. The example of Kepler gives us grounds to believe that the answers to ecological and religious questions can equally be found in Plato as in studies by climatologists and theologians.

How the Dialogue with other religious traditions affected my thinking, what I learned from it, what new methodological perspective aimed at coping with the ecological crisis can be inspired—if any—by this experience?

This question seems to presuppose that theologians work in isolated ivory towers. The ecumenical and interreligious movement is one of the features of contemporary theology. I would even say the opposite—it is hard to imagine theological thought today not being affected by other religious traditions, even if only by acknowledging their existence. In the context of the ecological crisis we must be more than just aware—we must establish a dialogue paving the way to thoroughgoing cooperation. As far as I am concerned, the work on the origins of Christianity inevitably involves dialogue with other religious and philosophical traditions. It would be impossible to understand Christian theology without relating to the Hebrew, Greek, and Roman Mediterranean religions, but also to the Eastern religions, Gnostics, and various currents of spirituality.

In fact, my previous experience made it easier for me to take part in the Dialogue. I felt quite comfortable in the exchanges on religious subjects. But one major question remains: is theology (or must it be) a confessional discipline? One is never only a "theologian." A theologian always represents a community—Christian or Muslim; Catholic, Lutheran, Greek Orthodox, Russian Orthodox, and so on. Hence the twofold concern to express yourself in the name of your own community and to make yourself understood by all the other communities. This is an extremely tricky task but the urgency of the ecological crisis has driven each of us back to explore our own religious tradition to look for answers that will be both coherent with that tradition and can be understood by others. Today this methodological approach has emerged everywhere whenever religions are called on to work together.

16

Mobilizing Religions for the Ecological Crisis: An Urgent but Difficult Task

Ignazio Musu

The Dialogue confirmed that the transformative energies of religions can be mobilized to cope with the ecological crisis but the discussion suggested that this was a far from easy task. It depends on what we might call the "cosmological vision" underlying religion. This concept of cosmological vision includes the relationship of human beings with the "cosmos."

The Dialogue focused on monotheistic religions while also considering some aspects of indigenous traditions concerning religion and ecology. If we widen the perspective to other religions, however, such as Buddhism, Daoism, or Hinduism, we immediately realize the importance of cosmological vision.

In monotheistic religions the extension of "nature" to "creation" is in a way implicit. This is not the case in other religions in which the definition of a role of the human being in the cosmos is not so evident. They see humans as a part of nature, the entire universe as God, while reincarnation confirms a continuity between different forms of human lives.

In Eastern religions an attitude of respect toward nature easily ensues from these cosmological ideas. In monotheistic religions the problem is more complex. As clearly emerged in the Dialogue, two attitudes toward ecological challenges arise from two different outlooks.

The first sees the creation as a gift of God. The idea of preserving creation is thus a natural consequence; it implies a specific interpretation of the command of "stewardship." Man as a steward must respect and preserve the gift of God.

In the second outlook the concept of "gift" is interpreted in the spirit of the Parable of Talents: the gift is a talent that each person has the duty to invest for a return. This point of view immediately moves toward a more active role of human beings as regards the creation. The conflict between man as "master" of other creatures and man as "steward" is not so strong. But it does exist.

This was very clear during the Dialogue. Some insisted on the duty of preserving the gift of God, with implications for lifestyles and in some case with critical assessments of economic activity and particularly of economic growth. Others, on the contrary, insisted on the role of human beings in transforming nature; they attributed a great value to the role of inventive activity and to research and technology. Strong support for this point of view comes from an opportunity offered by Christianity: the resurrection and the human contribution in re-creating the "suffering cosmos" with the prospect of "new heavens and new earth."

The only possibility of reconciling these two points of view comes from the acceptance of the idea of "limit." Limit not only in the sense of recognizing the constraints imposed by the principles of thermodynamics but also in the sense of being humble enough to recognize the possibility that technology can make big mistakes and these mistakes may be revealed by the unexpected results from new technological choices.

Through the acknowledgment of the idea of limit, we can reconcile the duty of preserving the gift of creation when transforming it. An attempt at this kind of "reconciliation" was made during the Dialogue, although I am in no position to judge whether it was successful.

What this experience did confirm is that religions may not automatically be a positive force in helping humankind to deal with the ecological crisis. The possibility of addressing successfully the global challenges of the ecological crisis may lie in building a common set of moral values guiding the behavior of people and institutions in a more environmentally friendly direction. To make a contribution in this direction, religions must feel the need for reciprocal knowledge of their different cosmologies and also the need to find a common ground where the various implications of these cosmologies converge. This implies giving up sectarian positions. No one can claim to have the final decisive answer. And this is something very difficult for religions. To contribute in this perspective, a religion should recognize that its ultimate value may be achieved through different routes and joint research with other religions may be very productive.

To achieve this, the first requirement is open-mindedness in those speaking on behalf of one specific religion in the debate. This open-mindedness is even more necessary if there are—as was clear during the Dialogue—different ways of involving the same religion to address ecological problems. This is an extremely difficult operation. The Dialogue confirmed just how difficult it is. There is an urgent need to consolidate moral values that will become part of the beliefs of people and governments so as to inspire their practical behavior. It is important that the various religions preach the need for environmental responsibility. In too many areas, however, religious teaching continues only to be theoretical, consisting of preaching and declarations without any consistent action. What a religion's followers proclaim and what they do are often different. To become mobilizing forces, religions need to involve moral codes that will then inspire political action.

The experience at the Dialogue, also during the more difficult phases, confirmed that to provide a practical perspective so that religions may help in successfully addressing ecological challenges, efforts must be pursued with open-mindedness and humility. Only in this way will we arrive at a common ground where a set of shared moral values can act as a driving force for individual and "political" behaviors.

17

EVOLVE: MODERNIZATION AS THE ROAD TO SALVATION

*Michael Shellenberger and Ted Nordhaus**

Sometime around 2014, Italy will complete construction of 78 mobile floodgates aimed at protecting Venice's three inlets from the rising tides of the Adriatic Sea. The massive doors—20 meters by 30 meters, and 5 meters thick—will, most of the time, lie flat on the sandy seabed between the lagoon and the sea. But when a high tide is predicted, the doors will empty themselves of water and fill with compressed air, rising up on hinges to keep the Adriatic out of the city. Three locks will allow ships to move in and out of the lagoon while the gates are up. Nowhere else in the world have humans so constantly had to create and re-create their infrastructure in response to a changing natural environment than in Venice. The idea for the gates dates back to the 1966 flood, which inundated 100 percent of the city. Still, it took from 1970 to 2002 for the hydrologist Robert Frassetto and others to convince their fellow Italians to build them. Not everyone sees the oscillating and buoyant floodgates as Venice's salvation. After the project was approved, the head of World Wildlife Fund Italy said, "Today the city's destiny rests on a pretentious, costly, and environmentally harmful technological gamble." In truth, the grandeur that is Venice has always rested—quite literally—on a series of pretentious, costly, and environmentally harmful technological gambles. Her buildings rest upon pylons made of ancient larch and oak trees ripped from inland forests a thousand years ago. Over time, the pylons were petrified by the saltwater, infill was added, and cathedrals were constructed. Little by little, technology helped transform a town of humble fisherfolk into the city we know today. Saving Venice has meant creating Venice, not once, but many times since its founding. And that is why her rescue from the rising seas serves as an apt metaphor for solving this century's formidable environmental problems. Each new act of salvation will result in new unintended consequences, positive and negative, which will in turn require new acts of salvation. What we call "saving the Earth" will, in practice, require creating and re-creating it again and again for as long as humans inhabit it.

Many environmentally concerned people today view technology as an affront to the sacredness of nature, but our technologies have always been perfectly natural. Our animal skins, our fire, our farms, our windmills, our nuclear plants, and our solar panels—all 100 percent natural, drawn, as they are, from the raw materials of the Earth. Furthermore, over the course of human history, those technologies have not only been created by us, but have also helped create us. Recent archaeological evidence suggests that the reason for our modern hands, with their opposable thumbs and shorter fingers, is that they were better adapted for tool use. Ape hands are great for climbing trees but not, it turns out, for striking flint or making arrowheads. Those prehumans whose hands could best use tools gained an enormous advantage over those whose hands could not. As our hands and wrists changed, we increasingly walked upright, hunted, ate meat, and evolved. Our upright posture allowed us to chase down animals we had wounded with our weapons. Our long-distance running was aided by sweat glands replacing fur. The use of fire to cook meat allowed us to consume much larger amounts of protein, which allowed our heads to grow so large that some prehumans began delivering bigger-brained babies prematurely. Those babies, in turn, were able to survive because we were able to fashion still more tools, made from animal bladders and skins, to strap the helpless infants to their mothers' chests. Technology, in short, made us human. Of course, as our bodies, our brains, and our tools evolved, so too did our ability to radically modify our environment. We hunted mammoths and other species to extinction. We torched whole forests and savannas in order to flush prey and clear land for agriculture. And long before human emissions began to affect the climate, we had already shifted the albedo of the Earth by replacing many of the world's forests with cultivated agriculture. While our capabilities to alter our environment have, over the past century, expanded substantially, the trend is long-standing. The Earth of one hundred or two hundred or three hundred years ago was one that had already been profoundly shaped by human endeavor. None of this changes the reality and risks of the ecological crises humans have created. Global warming, deforestation, overfishing, and other human activities—if they do not threaten our very existence—certainly offer the possibility of misery for many hundreds of millions, if not billions, of humans and are rapidly transforming nonhuman nature at a pace not seen for many hundreds of millions of years. But the difference between the new ecological crises and the ways in which humans and even prehumans have shaped nonhuman nature for tens of thousands of years is one of scope and scale, not kind. Humans have long been cocreators of the environment they inhabit. Any proposal to fix environmental problems by turning away from technology risks worsening them, by attempting to deny the ongoing coevolution of humans and nature.

Nevertheless, elites in the West—who rely more heavily on technology than anyone else on the planet—insist that development and technology are the causes of ecological problems but not their solution. They claim that economic sacrifice is the answer, while living amid historic levels of affluence and abundance. They consume resources on a vast scale, overwhelming whatever meager conservations they may partake in through living in dense (and often fashionable) urban enclaves, driving fuel-efficient automobiles, and purchasing locally grown produce. Indeed, the most visible and common expressions of faith in ecological salvation are new forms of consumption. Green products and services—the Toyota Prius, the efficient washer/dryer, the LEED—certified office building—are consciously identified by consumers as things they do to express their higher moral status. The same is true at the political level, as world leaders, to the cheers of the left-leaning postmaterial constituencies that increasingly hold the balance of political power in many developed economies, offer promise after promise to address climate change, species extinction, deforestation, and global poverty, all while studiously avoiding any action that might impose real cost or sacrifice upon their constituents. While it has been convenient for many sympathetic observers to chalk up the failure of such efforts to corporate greed, corruption, and political cowardice, the reality is that the entire postmaterial project is, confoundingly, built upon a foundation of affluence and material consumption that would be considerably threatened by any serious effort to address the ecological crises through substantially downscaling economic activity. It's not too difficult to understand how this hypocrisy has come to infiltrate such a seemingly well-meaning swath of our culture. As large populations in the developed North achieved unprecedented economic security, affluence, and freedom, the project that had centrally occupied humanity for thousands of years—emancipating ourselves from nature, tribalism, peonage, and poverty—was subsumed by the need to manage the unintended consequences of modernization itself, from local pollution to nuclear proliferation to global warming. Increasingly skeptical of capitalist meritocracy and economic criteria as the implicit standards of success at the individual level and the defining measure of progress at the societal level, the post-Second World War generations have redefined normative notions of well-being and quality of life in developed societies. Humanitarianism and environmentalism have become the dominant social movements, bringing environmental protection, preservation of quality of life, and other "life-political" issues, in the words of British sociologist Anthony Giddens, to the fore. The rise of the knowledge economy—encompassing medicine, law, finance, media, real estate, marketing, and the nonprofit sector—has further accelerated the West's growing disenchantment with modern life, especially among the educated elite. Knowledge workers are more alienated from the products of their labor than any other class in history, unable to claim some role in producing food, shelter, or even basic consumer products. And yet they can afford to spend time in beautiful places—in their gardens, in the countryside, on beaches, and near old-growth forests. As they survey these landscapes, they tell themselves that the best things in life are free, even though they have consumed mightily to travel to places where they feel peaceful, calm, and far from the worries of the modern world. These postmaterial values have given rise to a secular and largely inchoate eco-theology, complete with apocalyptic fears of ecological collapse, disenchanting notions of living in a fallen world, and the growing conviction that some kind of collective sacrifice is needed to avoid the end of the world. Alongside those dark incantations shine nostalgic visions of a transcendent future in which humans might, once again, live in harmony with nature through a return to small-scale agriculture, or even to hunter-gatherer life. The contradictions between the world as it is-filled with the unintended consequences of our actions—and the world as so many of us would like it to be result in a pseudorejection of modernity, a kind of postmaterialist nihilism. Empty gestures are the defining sacraments of this eco-theology. The belief that we must radically curtail our consumption in order to survive as a civilization is no impediment to elites paying for private university educations, frequent jet travel, and iPads. Thus, eco-theology, like all dominant religious narratives, serves the dominant forms of social and economic organization in which it is embedded. Catholicism valorized poverty, social hierarchy, and agrarianism for the masses in feudal societies that lived and worked the land. Protestantism valorized industriousness, capital accumulation, and individuation among the rising merchant classes of early capitalist societies and would define the social norms of modernizing industrial societies. Today's secular eco-theology values creativity, imagination, and leisure over the work ethic, productivity, and efficiency in societies that increasingly prosper from their knowledge economies while outsourcing crude, industrial production of goods to developing societies. Living amid unprecedented levels of wealth and security, ecological elites reject economic growth as a measure of well-being, tell cautionary tales about modernity and technology, and warn of overpopulation abroad now that the societies in which they live are wealthy and their populations are no longer growing. Such hypocrisy has rarely been a hindrance to religion and, indeed, contributes to its power. One of the most enduring characteristics of human civilization is the way ruling elites espouse beliefs radically at odds with their own behaviors. The ancient Greeks recited the cautionary tales of Prometheus and Icarus while using fire, dreaming of flight, and pursuing technological frontiers. Early agriculturalists told the story of the fall from Eden as a cautionary tale against the very agriculture they practiced. European Christians espoused poverty and peacemaking while accumulating wealth and waging war. In preaching antimodernity while living as moderns, ecological elites affirm their status at the top of the postindustrial knowledge hierarchy. Affluent developed-world elites offer both their less well-to-do countrymen and the global poor a laundry list of don'ts—don't develop like we developed, don't drive tacky SUVs, don't overconsume—that engender resentment, not emulation, from fellow citizens at home and abroad. That the ecological elites hold themselves to a different standard while insisting that all are equal is yet another demonstration of their higher status, for they are thus unaccountable even to reality. Though it poses as a solution, today's nihilistic eco-theology is actually a significant obstacle to dealing with ecological problems created by modernization—one that must be replaced by a new, creative, and life-affirming worldview. After all, human development, wealth, and technology liberated us from hunger, deprivation, and insecurity; now they must be considered essential to overcoming ecological risks.

There's no question that humans are radically remaking the Earth, but fears of ecological apocalypse—of condemning this world to fiery destruction—are unsupported by the sciences. Global warming may bring worsening disasters and disruptions to rainfall, snowmelts, and agriculture, but there is little evidence to suggest it will deliver the end of modernization. Even the most catastrophic United Nations scenarios predict rising economic growth. While wealthy environmentalists claim to be especially worried about the impact of global warming on the poor, it is rapid, not retarded, development that is most likely to protect the poor against natural disasters and agricultural losses. What modernization may threaten most is not human civilization, but the survival of those nonhuman species and environments we care about. While global warming dominates ecological discourse, the greatest threats to nonhumans remain our direct changes to the land and the seas. The world's great, diverse, and ancient forests are being converted to tree plantations, farms, and ranches. Humans are causing massive, unprecedented extinctions on Earth due to habitat destruction. We are on the verge of losing primates in the wild. We have so overfished the oceans that most of the big fish are gone. The apocalyptic vision of eco-theology warns that degrading nonhuman natures will undermine the basis for human civilization, but history has shown the opposite: the degradation of nonhuman environments has made us rich. We have become rather adept at transferring the wealth and diversity of nonhuman environments into human ones. The solution to the unintended consequences of modernity is, and has always been, more modernity—just as the solution to the unintended consequences of our technologies has always been more technology. The Y2K computer bug was fixed by better computer programming, not by going back to typewriters. The ozone-hole crisis was averted not by an end to air conditioning but rather by more advanced, less environmentally harmful technologies. The question for humanity, then, is not whether humans and our civilizations will survive, but rather what kind of a planet we will inhabit. Would we like a planet with wild primates, old-growth forests, a living ocean, and modest rather than extreme temperature increases? Of course we would—virtually everybody would. Only continued modernization and technological innovation can make such a world possible. Putting faith in modernization will require a new secular theology consistent with the reality of human creation and life on Earth, not with some imagined dystopia or utopia. It will require a worldview that sees technology as humane and sacred, rather than inhumane and profane. It will require replacing the antiquated notion that human development is antithetical to the preservation of nature with the view that modernization is the key to saving it. Let's call this "modernization theology." Where eco-theology imagines that our ecological problems are the consequence of human violations of a separate "nature," modernization theology views environmental problems as an inevitable part of life on Earth. Where the last generation of ecologists saw a natural harmony in Creation, the new ecologists see constant change. Where ecotheologians suggest that the unintended consequences of human development might be avoidable, proponents of modernization view them as inevitable, and positive as often as negative. And where the ecological elites see the powers of humankind as the enemy of Creation, the modernists acknowledge them as central to its salvation. Modernization theology should thus be grounded in a sense of profound gratitude to Creation—human and nonhuman. It should celebrate, not desecrate, the technologies that led our prehuman ancestors to evolve. Our experience of transcendence in the outdoors should translate into the desire for all humans to benefit from the fruits of modernization and be able to experience similar transcendence. Our valorization of creativity should lead us to care for our cocreation of the planet.

The risks now faced by humanity are increasingly ones of our own making—and ones over which we have only partial, tentative, and temporary control. Various kinds of liberation—from hard agricultural labor and high infant mortality rates to tuberculosis and oppressive traditional values—bring all kinds of new problems, from global warming and obesity to alienation and depression. These new problems will largely be better than the old ones, in the way that obesity is a better problem than hunger, and living in a hotter world is a better problem than living in one without electricity. But they are serious problems nonetheless. The good news is that we already have many nascent, promising technologies to overcome ecological problems. Stabilizing greenhouse gas emissions will require a new generation of nuclear power plants to cheaply replace coal plants as well as, perhaps, to pull carbon dioxide out of the atmosphere and power desalination plants to irrigate and grow forests in today's deserts. Pulling frontier agriculture back from forests will require massively increasing agricultural yields through genetic engineering. Replacing environmentally degrading cattle ranching may require growing meat in laboratories, which will gradually be viewed as less repulsive than today's cruel and deadly methods of meat production. And the solution to the species extinction problem will involve creating new habitats and new organisms, perhaps from the DNA of previously extinct ones. In attempting to solve these problems, we will inevitably create new ones. One common objection to technology and development is that they will bring unintended consequences, but life on Earth has always been a story of unintended consequences. The Venice floodgates offer a pointed illustration. Concerns raised by the environmental community that the floodgates would impact marine life have been borne out—only not in the way they had feared. Though the gates are still under construction, marine biologists have announced that they have already become host to many coral and fish species, some of which used to be found only in the southern Mediterranean or Red Sea. Other critics of the gates have questioned what will happen if global warming should raise sea levels higher than the tops of the gates. If this should become inevitable, it is unlikely that Venetians would abandon their city. Instead, they may attempt to raise it. One sweetly ironic proposal would levitate the city by blowing carbon dioxide emissions two thousand feet below the lagoon floor. Some may call such strong faith in the technological fix an instance of hubris, but others will simply call it compassion.

The French anthropologist Bruno Latour has some interesting thoughts on the matter. According to Latour, Mary Shelley's Frankenstein is not a cautionary tale against hubris, but rather a cautionary tale against irrational fears of imperfection. Dr. Frankenstein is an antihero not because he created life, but rather because he fled in horror when he mistook his creation for a monster—a self-fulfilling prophecy. The moral of the story, where saving the planet is concerned, is that we should treat our technological creations as we would treat our children, with care and love, lest our abandonment of them turn them into monsters. "The sin is not to wish to have dominion over nature," Latour writes, "but to believe that this dominion means emancipation and not attachment." In other words, the term "ecological hubris" should not be used to describe the human desire to remake the world, but rather the faith that we can end the cycle of creation and destruction.

NOTE

* This text was written after the San Giorgio Dialogue and published in the Fall/Winter 2011 issues of *Breakthrough Journal.*

18

MY MANIFESTO FOR THE DIALOGUE

Bruno Latour

Assembled in Venice, the most beautiful, most technology dependent, most threatened creation of human audacity and ingenuity, we wish to bear testimony that there exist other ways to handle ecological conflicts. Gathered for three days in the island of San Giorgio at the initiative of Giorgio Cini Foundation, our self-assembled group of ecologists, theologians, anthropologists, and social scientists, offer the following testimony encouraging other participants around the world to dismiss, discuss, or amend their propositions.

It is our conviction that the diversity and magnitude of conflicts over natural resource offers a unique opportunity to rejuvenate the links between science and technology, the various traditions of political ecology but also religion. Instead of claiming that it is possible and desirable to separate ecology and religion as much as possible, or to condemn ecological movements because it would risk becoming a "new religion," we wish to tackle head on the question of ecology and religion and offer an alternative description of their entanglement. Even though bringing religious passions to bear on ecological conflicts might be risky, we think the risk is worth taking because religious traditions might have the intellectual, ritual, and emotional resources at the scale of the challenge faced by the human-transformed cosmos.

We wish to start from the following conjectures:

1. religions have always been, for the best and for the worst, deeply concerned with the right ways to bless, transform, manage, and upkeep the cosmos; thus there is no sense in separating inside religious traditions that concern their rituals and theologies from their attitudes toward the cosmos and the consequences of their mundane organizations; in practice religions have always engaged in some form of eco-theology;
2. sciences in their Western traditions and from the seventeenth century onward have always been concerned with the right ways to understand

the connections between natural, political, and spiritual agencies so that in practice they can be defined as a enterprise in "natural theology," which has never been interrupted especially when they claim to have broken away from their irrational past;

3. ecological movements, although connected in many ways with religious traditions and scientific results, because of their relative novelty, have not yet found the right ways to harness the transformative power of religions nor to respect the innovative capacities of science and technology;

4. the result of those three conjectures is that the notion of "nature" might not offer a common ground for settling the numerous conflicts we have collectively to tackle, nor to understand the way other civilizations have lived in the world; it is thus our conviction that we should not engage in saving or protecting "nature" but in finding an alternative to the very notiono f nature.

In sum, we think that it is just when we enter what is described as an "era of limits" that we should not limit ourselves to the boundaries between science and religion, nature and culture, protection and salvation that we have inherited from the first modernizations. It is because we have only one planet to live on, that we should break through the narrow confines of modernity.

It is our belief that a new set of transformations is offered to religious souls, to concerned scientists and to ecologically minded activists once those three elements are brought together: religions that accept their cosmic impact; sciences that accept their theological concerns; ecological movements that accept to be religious and scientific in new ways. It is in this crucible that all our former values might be thrown and resuscitated. It is at this historical juncture that they—religion, science, and ecology—may finally come down to earth.

That ecological conflicts offer unique opportunities to religions, especially Christian traditions, is clear. They allow them:

- to abandon "nature" and "naturalism" as what they have to be opposed to in order to define their own beliefs;
- to put into question their obsession with moral problems and their anthropocentrism disconnected from cosmic concerns;
- to renew their own notions of Incarnation and providence;
- to reconnect with their own history of creation and re-creation and harness again the transformative energy that has led them to transform the world so dramatically;
- to extract themselves out of this long hate/love relation with science that has forced them to migrate either into the supernatural or to the deep recess of the inner souls;
- at the occasion of their meetings with so many newly minted divinities like Gaia, they may test their own prejudices against "paganism," "immanence," "nature," "pantheism," and deeply renew what they mean by transcendence and spirituality.

In brief, it is possible that Christian religions may start all over again to provide rituals and energy to give a new meaning to what is meant by Incarnation and the long sacred history of the transformative power of the Holy Spirit.

But ecological conflicts are an immense chance for scientists and engineers as well because they will allow them:

- to abandon their pretence to have broken from religion, metaphysics, and politics in the name of a "nature" now too multiple and too composite to be unified once and for all;
- to complicate their quest for certainty, mastery, and indisputability that put them in the impossible position of settling conflicts that cannot be closed by them only;
- to reconnect with religions and thus reinterpret their long and beautiful history as a long quest for the right type of "natural theology";
- to allow themselves to wonder and worship at a cosmic adventures without having to pretend that they share "a view from nowhere" that they have to claim is valueless; the value of truth and objectivity deserves better;
- to unleash the new creative power of their trades without having to ignore unintended consequences through the invention of a new set of values that would tie the contradictory demands of audacity and care;
- at the occasion of their meeting with the new divinities of ecological movements (such as Gaia) they too can begin to test again what they meant by a nonhuman view that is no longer a view from "nowhere."

In brief, science and technology may begin all over again in the confidence that by shifting their values from the search from "matters of fact" to the handling of "matters of concern" they remain faithful to their original vocation and their final destiny.

But it is of course for ecological movements that the opportunities are just as great so that they may answer differently to their vocations. When encountering renewed religious, scientific, and technological practices, they might seize this occasion:

- to abandon "nature" as what should be protected or saved, as if it could provide a final judge to shortcut political conflicts;
- to escape from their insistence on limiting human footprints and withdrawing from the effects of human innovations just at the time when those effects have to be tackled with expanded ingenuity and innovation;
- to break out of the limits of cosmocentrism since the overwhelming transformations of the planet shift the burden back to human shoulders;
- to reconcile themselves with the importance of technical innovations and scientific research to re-create this "nature" that has been forever transformed;
- to participate in the renewal of engineering values in order to redefine on which ground technical innovations should be sorted and how to tend for the unintended consequences of our collective actions;

- to join forces with religious and especially Christian traditions not for the dominion over "nature," but for the common task of continuing Creation.

It is because neither God's will, nor the results of the various sciences, nor the conflicting states of a "nature" to be protected will ever be able to offer a final arbitration to define the common good, that it is necessary to explore other ways to move collectively through controversies. We are well aware that this way to shift the conversation between religions, sciences, and ecological movements remains highly tentative, but we are confident that they offer a different and more hopeful entry into the politics of environment.

19

Roman Catholic Commitment to Promoting Sustainability

Andrea Vicini

"I Have a Dream": A Role for Religions and for Catholics

Ted Nordhaus and Michael Shellenberger begin their book *Break Through* by referring to the historic speech by Martin Luther King "I have a dream."[1] They highlight its strength, its focus on what is positive and desirable, its ability to point us toward the future with a contagious hope and with a strong longing for freedom and justice. I profoundly love that speech. I read it often. I dream it too. It always moves me to tears. I used it in teaching and in homilies. I turn to it to nourish my hope in a better future. Such a future can come. We can play a significant role in making it happen, sooner than later, here and now.

It is this positive and inspiring dream in a just and sustainable future that led me to accept the invitation to our *Dialogue*. I am profoundly convinced that all religious traditions, together with all people of good will, can make such a positive contribution. In different ways, religions stimulate and gather human imagination, ingenuity, and creativity. Within each religious tradition, human values are nurtured. Just and virtuous practices are promoted and supported. Religious traditions help us to strengthen our role as ethical partners and players, as believers and as citizens in today's globalized world.

Religions played a transformative role in the past. They influenced cultures and civilizations. They shaped progress and urbanization. In some instances and during specific historic periods, religions had a positive prophetic role, indicating the greater justice that was needed. They helped to devolve human, ecclesial, and social energies to achieve it. Religions continue to be involved in this transformative and prophetic process. At the same time, conditions are needed to make it happen and to avoid less positive historic examples where religions showed a problematic and scandalous authoritarian, intolerant, oppressive, and violent side. To say it in other words, the ways in which religions aim at transforming reality matter. The outcome can be positive or negative. We want it to be the best possible, for believers and for all humankind alike.

A WAY OF PROCEEDING

What are these conditions? Before examining them, we need to look at the current context because these conditions are embedded in it. Hence, we need to focus on the Catholic Church and leave to others any consideration concerning their own religions.

In some countries around the world (e.g., in Western countries), Catholicism is challenged. Mostly, it is its institutional component that is under scrutiny, that is, what Avery Dulles called the "Church as institution" in his famous analysis of Church's models.[2] Spirituality, on the contrary, is strongly alive and well. It is defined as one's free way to relate to God, to interact with the Divine, to experience the Transcendent here and now, both individually and communally. In light of such significant spiritual dimension, Catholic leaderships and believers should verify their ability to be challenged and questioned by the growing number of people who feel uncomfortable within the Catholic Church, but less in the Catholic faith or within the larger Christian tradition.[3] The whole Catholic Church—all included—should learn how to question herself in light of this malaise among her believers. We should advance in revising our practices and priorities when we become aware that we need to rearticulate our faith in light of the challenges that come from human experience and from today's social transformations.

Beside this tension between institutional components and believers (including charismatic historic figures and theologians), when we consider sustainability within Catholicism, we notice how it squares well with original and foundational Christian and Catholic experiences concerning the relationship between God, creatures, and creation. We can confirm it by focusing on the Bible, but also on theological discourse throughout the centuries, and on the experience and practices of Catholic believers during history in all continents.

While we highlight this continuity, at the same time we need to consider discontinuities: renewal and change took place and are still occurring. Even when we read the Bible, we become more and more aware of how it challenges us and how it calls us to conversion. Our anthropocentric approach to sustainability needs to be reinterpreted in a more relational way, by avoiding a focus on mastering and dominating creation without respecting it responsibly. Likewise, the experiences of Catholic communities, with their struggles and achievements, invite us to see how Catholic official teaching, theological reflection, and concrete practices (within the Church and society at large) need to be discussed, verified, and maybe changed to make our commitment to sustainability coherent and effective.

This way of proceeding (that we could call "methodological approach" or "conditions of feasibility and progress") demands our commitment to be informed believers, to listen to others, to understand them with profound respect and care, to communicate what many believers and citizens are already doing to promote sustainability. Moreover, it invites us to reflect theologically in order to provide theological resources for practical engagement, for continuing our dialogue within society and with other religious traditions, and for putting our creative energies together. Within the Catholic community, prayer and celebration (e.g., para-liturgies) are further means to nourish and to strengthen our reverence[4] and greater appreciation for the gift of creation in order to protect it concretely and responsibly.

To Go Forward: Discernment and Accountability

Hence, discernment in decision making and in action becomes essential. It leads us to distinguish between common interest, public interest, and common good, where the common good indicates our struggle toward a greater global justice as an ongoing (and hopefully achievable) task, that will benefit all citizens and creatures—particularly the less privileged.

Discernment highlights how we should be attentive to the many positive ways and models in which we might promote sustainability, both with technological innovation and with personal and social sustainable choices in our lifestyle (e.g., in reducing our carbon footprint). To this emphasis on technological innovation (with the funding and investments that it requires) and to the focus on virtuous personal and social behavior, I add that, within the Catholic Church and within society at large, we should be accountable when we do not support sustainability: particularly, when we do not sufficiently encourage scientists and researchers to go forward and to propose new sustainable technologies; and, when we do not bother verifying how sustainable is our personal, institutional, and social lifestyle. Accountability aims at strengthening and at expanding our commitment to sustainability. Its goal is promoting a greater responsibility and a more effective solidarity within society at large and with all creation.

A Continuous Dialogue

During our Dialogue we experienced concretely the possibility and the benefit of interacting respectfully among nonbelievers and believers belonging to various religious traditions. We were challenged by any participant and we challenged them. We reflected critically on our own religious history and on the current involvement of each religious tradition. Mutual respect, willingness to know and to understand, critical appreciation, and desire of being part of a relationship shaped our Dialogue. This was humanly, ethically, and theologically extremely significant.

Personally, this concrete exercise strengthened my commitment to dialogue, to listen, and to be part of a conversation aimed at contributing to promote sustainability. It also confirmed the positive and needed role that religions, and religious people, can play and should play in addressing concrete matters like the sustainable conditions of life on our planet, for us and for future generations. Grateful for such an opportunity, I highlight its importance beyond the specific occasion of our gathering because it indicates a specific way of proceeding to address one of the major and most urgent issues in our contemporary world.

To the conversation between believers and unbelievers, Catholics can bring their longing for dialogue, collaboration, progress, justice, and their desire for trying to understand better what is good for each individual, for peoples, and for the whole humanity. When we reflect on dialogue, and when we are engaged in a conversation, we are well aware that dialogue requires clarity and commitment, with great humility and patience. We need to let ourselves be touched by the other's approach. We need to look at our own truths and to the truth of our own religious tradition by being opened to further developments, to insights that come from the other—individuals and diverse religious traditions as well. We search truth and we long for it, without exclusively owning it. It is from the standpoint of truth that I experience my religious tradition and that I engage myself in the long-life journey of searching for the many other sparks of truth that are well beyond my own faith, life, culture, and experience. Hence, the others' truths, welcomed and shared, become my own truth.

A Vision

Historically, we can identify a long series of models of interaction between religious traditions and civil society, between church and state. Without entering in this complex analysis, we need to be aware of this plurality of models and of concrete historic actuations when we reflect on the contribution to promote sustainability that can come from religious traditions. Religions, including Catholicism, can aspire to play a positive role within society without making society "religious." They can point to what we should strive to achieve. They are engaged in making it happen by collaborating with anyone, well beyond the ecclesial borders.

Moreover, in what concerns the environment, we clarify how the Catholic Church relates with society at large by defining the various models of eco-theo-politics that mark her long history and those that are operative today. We could do the same for each religious tradition. These models are multiple, sometimes concordant, quite often contrasting and opposed. They are part of the complexity of each religious tradition throughout human history. These various political visions result from different worldviews concerning the relationship between God and God's presence within the universe. The eco-theo-politics that we are longing for, and that we want to support, aims at defining, promoting, and strengthening just sustainability. We should not forget that sustainable choices and practices, both individual and collective, are not exclusively a Catholic prerogative. They express our commitment as citizens and our interest in our own well-being as well as in the well-being of creation and of future generations.

Concretely, our Dialogue motivates me further to continue in revising Catholic theological language by reinterpreting it and by balancing what in many instances appears as an outspoken anthropocentrism. A more moderated and enlightened approach, which highlights our responsibility and our stewardship for creation and within it, appears to be more promising. The goal is to consider all living creatures as part of creation by stressing relationships and interactions. At the same time, we should continue to search for better virtuous ways of articulating concretely our responsibility for creation by showing a greater respect for all living creatures.

Finally, as believers, we continue to read the Bible and to hear its call to conversion. While we strengthen our ethical commitment to promote sustainability, education stands as a priority. As educated believers and citizens we can advocate for respecting the environment and we can achieve stewardship. Hence, Catholic educational institutions all over the world— for example, colleges and universities—are a privileged context where both research and lifestyle could be strengthened further. Increasingly, they could play a major role in promoting sustainability.

NOTES

1. See: Ted Nordhaus and Michael Shellenberger, *Break Through: From the Death of Environmentalism to the Politics of Possibility* (Boston and New York: Houghton Mifflin, 2007); Martin Luther Jr. King, "I have a dream" (1963), http://www.mlkonline.net/dream.html

2. See: Avery Dulles, *Models of the Church*, expanded ed. (New York: Doubleday, 1991).

3. See: Peter Steinfels, "Further adrift: the American church's crisis of attrition," *Commonweal* 137, no. 18 (2010): 16–21; The Pew Forum on Religion & Public Life, "U.S. Religious Landscape Survey" (2010), http://religions.pewforum.org/

4. See: Pontifical Council for Justice and Peace, *Compendium of the Social Doctrine of the Church* (Washington, DC: US Conference of Catholic Bishops, 2005), § 487; Jame Schaefer, "Appreciating the beauty of Earth," *Theological Studies* 62, no. 1 (2001): 23–52; Jame Schaefer, *Theological Foundations for Environmental Ethics: Reconstructing Patristic and Medieval Concepts* (Washington, DC: Georgetown University Press, 2009).

20

The Balance of Nature. What Can Be Learned from the History of the Earth and Life

George Theokritoff

Prologue

A prologue gives me an opportunity to introduce myself. I am a scientist, specializing in Earth History and Paleontology, now retired. In recent years, I have become interested in the interface between science and Christianity, more specifically the Theory of Evolution and Genesis. I had the opportunity to attend the Dialogue organized by the Giorgio Cini Foundation, to present some ideas and to observe how scholars from various disciplines and intellectual loyalties were diagnosing the present ecological crisis and what they were suggesting in order to cope with it. A concept that was often directly or indirectly evoked was the concept of balance, both in terms of diagnosis and in terms of therapy: is there a natural order or balance that is going to be altered, due to an unprecedented anthropogenic role played by human beings? How and to what extent can human beings restore a balance that seems to many irremediably altered?

The editors of this book asked me to comment retrospectively on the issue of the "harmony" or "balance" of nature—as it was dealt with in the Dialogue—from the scientific point of view of a paleontologist, so helping the reader to embed the current ecological crisis in a broader time/space horizon.

The Balance of Nature

Even though the term "nature" is used in a number of different senses,[1] I suggest that it is useful as designating a subset under the rubric "Creation," human volitional activity being another subset. There are actors and forces in the world with which humans may interact, but for which they do not set the agenda. Regarding them as part of "creation" reopens the possibility that "nature" in this sense can offer us guidance; to a degree at least it reflects God's will and speaks to us of him. In the context of environmental crisis and questions about human use of the world, this obviously brings us to the vexed question of the "harmony" or "balance" of nature that I will look at from the scientific perspective. While such ideas are often used in glib or simplistic ways, it would equally be a mistake to persuade ourselves that the *only* "order" in the world is whatever we decide to impose on it.

Because for millennia, humanity inevitably experienced "nature" within a short time-frame, it was perceived as harmonious. Thus, the "balance of nature" came to be seen as a harmonious attribute of "nature" carrying the implication that there are "forces" in "nature" that constrain change within relatively narrow limits. In examining the concept of balance or harmony, we should bear in mind that humanity is dependent on "nature" for its physical needs and hence we could usefully and correctly consider the two, humanity and "nature" as constituting one community.

Here, a number of questions come to mind: What is meant by "balance"? Is it interdependence? Is it stasis? Is it a paradisiacal state such as that portrayed in *The Peaceable Kingdom?*[2] Is it an emergent property within the humanity/"nature" community? To a significant extent, answers depend on the time-frame adopted.

In a short time-frame spanning decades, approximating the biblical three-score-and-ten years of a human life-span, the overarching theme is one of stasis. The daily and seasonal cycles do not differ very much from their predecessors. However, there are occasional intrusions: earthquakes, tsunami, volcanic eruptions, landslides, fires, and so on any of which can have dramatic consequences. But the effects are constrained, localized (although volcanic eruptions may add greenhouse gases, aerosols, and ash to the atmosphere on a global scale).

Still within the same timescale, there is a change that has directionality: seral succession, a succession of plant communities that is tracked by a succession of associated animal communities. Seral succession is not evolutionary change—it is ecological succession. Recovery from a catastrophic event involves an ordered succession of this sort, as the following example (from the southeastern United States) illustrates.

A fire in a forest exposes the soil to open sunlight. The first plants to colonize are those that are adapted to high light intensity: annual and perennial herbs and grasses that constitute *the pioneer community*. Over the next few years, pine seeds are blown in from surrounding areas and pine seedlings are established. As they grow, they compete with the herbs of the pioneer community for moisture and their expanding canopies shade the area, altering the microenvironment. By the time a stand of tall pines is in place, there are many oak and hickory seedlings on the forest floor but few pine seedlings. This is because oak and hickory seedlings prosper under shade conditions but pine seedlings do poorly. Later, the oak and hickory establish a well-developed understory under the pines. When a pine tree dies and falls, it opens the upper-story and is replaced by one of the oak-hickory community.

Eventually, almost all the pines are replaced by members of the oak-hickory community. This community is the *climax community*, a stable community adapted to the macroenvironment obtaining in that place at that time.

Fire is an environmental disturbance and we tend to see it as a destructive agent. It destroys timber that carries a monetary value and it even burns mega mansions in California. But it is also a creative agent, in that it opens up habitats suitable for the plants and associated animals of the pioneer community. It restarts the whole seral succession, giving everything an opportunity, thus maintaining biological diversity.

It is the most diversified ecosystems in which all niches are occupied and all resources are utilized that are the most stable and the most efficient. A relatively simple example may clarify this matter. In the northeastern United States and eastern Canada, the top predators (wolves, bobcats, and panthers) have been eliminated by human agency. The loss of these predators has resulted in a substantial increase in whitetailed deer populations. This burgeoning population causes extensive damage to the forests covering much of the region—they browse on the lower parts of trees and destroy much of the lower story vegetation such as shrubs and tree-seedlings. The new top predators, human hunters and motor vehicles, fall far short of the ongoing culling of the deer population achieved by the earlier top predators. Here the browse used by the deer is overutilized beyond the point of sustainability and the deer are underutilized by the predators. The elimination of tree-seedlings by the deer will eventually result in the collapse of the forest and the community dependent on it.

Should dwindling biodiversity worry us? Or is salvation to be sought in a "neo-nature"? If the entire 4.6 billion year history of the Earth were to be represented by 24 hours, humanity appeared in the last three seconds. In these last three seconds, humanity coevolved with its environment. What this means is that humanity is adapted to the present, albeit rapidly changing, environment.

At present, humanity occupies a wide ecological niche. This has been possible because humanity has created a number of microenvironments that maintain the appropriate temperature range, the availability of food and water, as well as other resources. But is it reasonable to extrapolate from this and suggest that in the event of the collapse of ecosystems globally, they can be replaced by new, technologically created ecosystems consisting of genetically engineered organisms replacing those that are extinct? One can only speculate on the outcome but a serious problem arises: it is easy to assume that it would be possible to bioengineer different kinds of neoorganisms in numbers to "replace" those that are extinct. But the crucial question is what it is that these neoorganisms would *do* in a new set of global environments. In other words, how can we be sure in advance how the planned neoorganisms would perform their intended *functions* in their neoenvironment? Would we suddenly discover that we have to bioengineer humanity as well?

A clear and highly relevant example of what could be called "balance of nature" comes to hand in the forces acting in opposition to each other in the shaping of climate. It is therefore appropriate to discuss some of the feedback pathways acting now.

An example of positive feedback readily comes to mind. As the oceans warm, evaporation increases, in turn increasing the amount of water vapor in the atmosphere. Because water vapor is a potent greenhouse gas, the opacity of the atmosphere to the reradiated infrared is increased. Furthermore, because the oceans redistribute heat globally, polar and glacial ice is attenuated, thus reducing the reflectivity of these regions and exposing rock surfaces that absorb heat. This causes the polar and glaciated regions to be converted from reflectors of heat to absorbers of heat, accelerating the melting of polar and glacial ice.

Similarly, examples of negative feedback readily come to mind. The increase in water vapor in the atmosphere can be expected to increase cloud-cover that in turn would result in the reflection of some solar radiation. In addition, there are carbon sinks. The oceans absorb carbon dioxide that is utilized in photosynthesis by phytoplankton. As the phytoplankton dies off, carbon is sequestered as it is buried in sediment on the ocean floor. Rock weathering, particularly of rocks rich in calcium silicates is another carbon sink. Here, the interaction of such rocks with carbon dioxide gives calcium carbonate that can be deposited as limestone or, in the case of acidified water, remain in solution as calcium bicarbonate. Although volcanoes emit carbon dioxide they also emit sulfate aerosols that reflect sunlight out into space. Our problem today is that the available negative feedback pathways are overwhelmed by the sheer rate at which carbon dioxide is added to the atmosphere.

Considered in the context of the geological timescale, a perspective emerges that is rather different from the sort of succession we might see within a human lifetime. And if we focus our discourse on greenhouses, icehouses, and extinctions, we can understand some of the major changes observed as the biosphere's response to environmental change. Some groups of organisms adapt sufficiently quickly and survive. Others fail to do so and become extinct.

Whether there is a "balance of nature" depends almost entirely on the time-frame chosen. In the case of the short time-frame, we are looking at a thin time-slice. We do not see dramatic changes (except as noted above) nor do the animals, enjoying normal life-spans, "know" that in a number of years, their "tribe" would be extinct. In the case of a long time-frame, such as that of geological time, we become more aware of changes of much larger latitude and duration (e.g., major extinction events, reconfiguration of continental distribution by Plate Tectonics). But such mega-events occur within a stack of thin time-slices—seen from within each thin time-slice, they may be just events, rather than mega-events.

If there is an answer to the question, "is there a balance of 'nature'?" it has to recognize that there are forces acting in opposition that can be tilted one way or another. The answer, then, is not *Yes* or *No*. It is *Yes* as well as *No*, with a porous boundary between the two.

It is at this point that we might ask how it is that science can help us. The matter gets confused because it is all too common that some consider that science is telling us what to do. "What to do" is beyond the remit of science. But science can tell us what is happening and advise us as to which actions are desirable and which are not. Nevertheless, whatever policies are finally adopted should be *informed* by science. One of the difficulties standing in the way of agreement on policy is that it is difficult to conceive of a world that is different from the present. This is not denial but by extension, it begets the conviction that there is no way that the world could possibly be different.

We are now ready to ask what it is that we can learn from the History of the Earth and of Life, with particular reference to greenhouse episodes and biological extinctions.

A Brief History of Greenhouses and Icehouses

The Earth has experienced a number of greenhouse episodes punctuated by icehouse episodes (see Table 20.1). Evidence for this is found in marine and lacustrine sediments, the distribution of fossils, the composition of gas bubbles and oxygen isotope ratios in ice sheets, and in other features of the geological record.

There were at least two or three glaciations in the Proterozoic Eon (2500–550MA [million years before present]). The last of these, the Late Proterozoic glaciation (900–600MA), has been described as worldwide or almost so (and has inspired the description "Snowball Earth") followed by warming. But this warming is not associated with anything like a major extinction. Instead, there is the initiation of an animal fossil record, one showing rapid diversification with the concomitant occupation of an expanding spectrum of ecological niches. In our present context, it is clear that the shape of the Late Proterozoic event differs markedly from that of our present event.

Although the Earth was probably almost ice-free only in the Mesozoic Era (245–66MA—the "Age of Dinosaurs"), there was a notable glaciation somewhat earlier, in the Permo-Carboniferous (270–255MA). The rapid warming event that ended this glaciation is associated with the biggest extinction event of all time. We shall take a closer look at this one presently.

Although the end-Cretaceous event, the one that saw the demise of the dinosaurs, is associated with moderate cooling and is not larger than earlier events (see Table 20.1). I simply note it, on account of the notoriety it has gathered.

The later, Pliocene-Pleistocene glaciation (colloquially called "the Ice Age," one populated by Neanderthals and Mammoths), as well as the end-Permian glaciation shows that icehouse events are not monolithic. Rather they are made up of a number of icehouses alternating with warmer interglacials, episodes that do not qualify as greenhouse episodes. We are currently in such an interglacial. In this connection, it is important to recall that until relatively recently it was thought that the earth was getting cooler. Evidence

Table 20.1 The Geological Timescale with summary of climate changes and major extinction events

MA	Geological Time-Units	Events
	HOLOCENE	Increase of CO_2 — reversal of global cooling and start of global warming
0.01		
2	PLEISTOCENE	Wide glacial/interglacial sea-level changes
		Ice-sheets established in northern hemisphere
	PLIOCENE	
5		
	MIOCENE	Antarctic ice-sheet established
24		
	OLIGOCENE	Start of long cooling—decrease of CO_2
37		Rapid warming—Global average 6–7° C warmer than present-breakdown of methane hydrates
	EOCENE	Cooling climates with increasing seasonality
58	PALEOCENE	
66		**MAJOR EXTINCTION** (End Cretaceous)
	CRETACEOUS	Warm climates with increasing latitudinal zonation
144		
208	JURASSIC	End-Triassic extinctions
	TRIASSIC	
245		
	PERMIAN	**MAJOR EXTINCTION (END PERMIAN)** **Extensive volcanism and warming** **Permo-Carboniferous glaciation**
286		
	CARBONIFEROUS	
360		
	DEVONIAN	Extinctions within Devonian
408		
	SILURIAN	End—Ordovician extinctions
438		
	ORDOVICIAN	End—Cambrian extinctions
505		
	CAMBRIAN	
570		
	PROTEROZOIC (900–600M A)	Very widespread glaciation ("Snowball Earth")
2500		Archean-Proterozoic transition
c. 4.8 billion (American usage)		Origin of Earth

Note: (1) In accordance with accepted geological practice, the earlier events are shown at the bottom and the latest at the top. Such tables are to be read from the bottom up if a chronological context is desired. (2) Please note the sequence of events leading to the end-Permian extinction: Glaciation ended by warming associated with the end-Permian extinction event. (3) MA = Millions of years before present.

was based on the vertical distribution of spores and pollen in the sediments deposited in postglacial lakes, oxygen isotope ratios in foraminifera vertically distributed in oceanic sediments, and various proxies studied in ice-cores. The science was sound but more recently, a global warming signal has emerged. The cooling trend has reversed and the scientific consensus identifies a significant anthropogenic factor.

The present event, even though it is in its early stages, is following the pattern of the end-Permian event. The very large excess of carbon dioxide added to the atmosphere is driven by the burning of fossil fuels, cement manufacture, agriculture, as well as "natural" sources such as volcanoes. This excess is well beyond the capacity of the available carbon sinks. If we were to suggest, with due caution, a plausible scenario as to how the present event might play out, it would predict that the warming event may last as long as 100,000 years if the end-Permian is taken as a model. But the duration of the warming would depend on the amount of carbon dioxide and other greenhouse gases added to the atmosphere from all sources. It is probable that fossil fuels would be exhausted within two centuries, given present rates of consumption. Even if the burning of fossil fuels be taken as the only significant source of greenhouse gases, warming would continue well after fossil fuels are exhausted because there is a great deal of inertia in the climate system. But within a more realistic framework, methane and water vapor are potent greenhouse gases too and their release to the atmosphere is triggered by warming. It is reasonable to expect that the increased addition of these gases would increase the *rate* of warming. We would then be heading for what has been described as an irreversible, runaway greenhouse.

A BRIEF HISTORY OF THE BIGGEST EXTINCTION OF ALL

Oscillations between greenhouse and icehouse states have a substantial effect on the biosphere. In this connection, our attention is currently drawn to the disappearance of whole swathes of organisms and many consider that we are now beginning to see a mass extinction. As in all extinction events, there is an overall loss of diversity. It is not so much the frequency (the number of individuals), but rather the diversity (number of kinds) that is the issue.[3]

But extinction is also part of the evolutionary process and it is going on all the time. In "normal" times, the rates of replacement more or less balance the background extinction rates. At the present time, the background extinction rate is strongly reinforced by habitat loss, global warming (global warming, however caused, also causes habitat loss), toxins, contractions of genetic pools driven by the abnormally high rate of environmental change, whether anthropogenic or "natural."

In order to recover any useful information from the geological record that might help us understand the present crisis, we need to gain an overview of the current crisis. Stated in stark terms, it is a warming event caused by the addition of large quantities of greenhouse gases to the atmosphere. The warming (climate change) is destabilizing the biosphere globally, as testified by the already rapid rate of extinctions. The relevance of a stable biosphere to survival of humanity has already been discussed.

If this is a fair assessment of the overall shape of present reality, we need to look particularly at those events in the history of the earth that show a similar pattern. In this connection, it should be noted that not all extinction events fall within this pattern, even though climate change may have been a significant additional factor in some.

In addition to a number of minor extinction events, the fossil record shows a smaller number of major extinction events (see Table 20.1). The most famous is the one that saw the demise of the dinosaurs. Rather than being associated with warming, this extinction seems to have taken place within a moderate cooling event, one continuing into the Paleocene. It is even not larger than earlier extinction events. I have included it on account of the notoriety it has gathered—it was caused, so it is claimed, by a bolide striking the earth. Sensational and hence well worthy of media attention, it has entered the folklore of our time. But paleontological evidence suggests that the Chicxulub crater (Yucatan Peninsula), claimed to be the smoking gun, is 300,000 years too early.

Regardless, the end-Cretaceous event that saw the extinction of the dinosaurs is not the biggest extinction of all time. That distinction belongs to the end-Permian event. Estimates vary but recent work suggests that 90 percent of all species and more than 95 percent of marine species did not survive this event. Some have suggested that all life came close to being snuffed out. Despite these very high numbers, lending this event a dramatic aura, it appears that the extinction event was spread over some 100,000 years. But recovery of the biosphere may have taken as long as 30 million years, an interval that would take us halfway back to the extinction of the dinosaurs.

What caused this event is not fully understood. There were probably many causative factors but the prime suspect is the enormous volcanic activity in Siberia that left volcanic rocks over an area of some 1.5×10^6 square km, ranging in thickness from 400 to 3,000 meters. The concomitant addition of enormous quantities of carbon dioxide and other gases ended the Permo-Carboniferous icehouse and ushered in the warm climates of the Mesozoic. This warming event probably triggered the release of large quantities of heat-trapping methane from the sediments accumulated on the continental shelves. In addition, warming of the oceans resulted in increased evaporation and the concomitant increase of atmospheric water vapor, itself a potent greenhouse gas, in the atmosphere. It should be noted that this warming event is associated with the end-Permian extinctions.

The above discussion suggests that it is the end-Permian event that resembles closely the present warming perhaps even in terms of scale and rate of change; it may well be that the end-Permian warming/extinction event is its closest precedent. But there are also differences. Obviously, there was no anthropogenic component. And at the present time, the quantity of carbon dioxide released by volcanoes is minor; it is approximately equal to 1 percent of the total carbon dioxide from the burning of fossil fuels.

Nevertheless, the parallels between the end-Permian warming and the present rapid warming strongly suggest that the biosphere (of which we are inextricably a part) is at serious risk. On the global scale, perhaps the most serious concern is that, as we have seen above, impoverished ecosystems are not stable. Although estimates differ, recovery of diversity in the biosphere could take as long as 30 million years, an interval that would take us halfway back to the extinction of the dinosaurs.

"RELIGIOUS" PASSIONS

The preliminary material distributed to participants to this dialogue called for consideration of the *kind* of passions, drive, commitment that do not emerge out of a religious matrix yet are comparable in degree with passions arising out of a religious matrix.

At least one example can be cited from recent history. During the Second World War, Winston Churchill was able to mobilize such "passions," partly by his oratorical skill (John Kennedy said of him "He mobilized the English language and sent it into battle") and partly by the historical context. In 1940, following the German military successes in western and northern Europe, the enemy was on the other side of a narrow strip of water, the English Channel, bent on the invasion of Britain. The entire population of Britain clearly understood that it was "looking down the barrel of the gun." Churchill warned that there would be sacrifices, recalled in his famous "I have nothing to offer but blood, tears, toil and sweat."[4] The force of this Churchillian "passion" had an interesting postwar manifestation. Many people in Britain considered that what they had done during the War, whether in the armed forces, the Home Front or elsewhere, was the most meaningful thing they had done in their lives.

In the present case of Global warming, the metaphorical gun does not exist in the minds of those who reject any anthropogenic role. But to many of those who accept the fact that there is Global warming, it all seems to belong to a distant future, that it might make some difference of unspecified kind by the end of the century.

NOTES

1. Celia Deane-Drummond, "Discerning creation in a scattering world: questions and possibilities," LEST VII Louvain Explorations in Systematic Theology, October 31, 2009.
2. A painting by Edward Hicks, a nineteenth-century American painter.
3. What is being counted in an extinction event to determine net loss? Not individuals because extinction events commonly result in enormous numbers of individuals (frequency) distributed among a reduced number of species (diversity). Genera and families are usually counted because there is greater stability in their concept

than in that of species. Above the familial level there is usually too great a survivability to give a reasonable understanding of a given extinction event. This will become clearer once we recall that if all the species assigned to a given genus go extinct, the genus itself goes extinct. One surviving species, perhaps out of several, ensures that the genus to which it is assigned remains extant. So it is for families: if one genus, perhaps out of several, remains extant, even if it is monospecific, the family remains extant.

4. WSC, May 13, 1940.

21

What Is My Personal Outcome from This Dialogue?

Eric Geoffroy

The very title of our Dialogue as well as the assumptions and statements in some of the papers reveal a dichotomy or even conflict between religion and science. This contrasts with the Islamic point of view that I was called to illustrate here. The Koran often urges man to search for "signs" both in the world and in himself, in scientific experimentation and in inner experience. That is why in our own time, the *ulama* and the Muslim scholars address ecology by referring to the famous *Sharia*—so misunderstood by many Muslims and non-Muslims—and by reviving the ancient science of "the ultimate aims of the *Shari'a*" (I will no longer spell it according to the popular media form). In actual fact the *Shari'a* is the cosmic law; and *mutatis mutandis* the equivalent of the Hindu term *Dharma*. This science enunciates five principles whereby all theology in Islam must aim at the well-being of humanity and all the other realms of the creation. The first principle is absolute respect for life, in all its forms.

Personally I feel this reviving of an ancient science is not sufficient because it is still a horizontal-type reform. In any case, it echoes the idea put forward by Bruno Latour that Christianity may be revitalized by the ecological crisis. I would say that only a spiritual revolution, both personal and collective, would seem to be able to meet the challenge. The Islamic liberation theology would appear to be more relevant: it is radical in the sense that it aims at the worship of the One, and aims to free man from all kinds of idolatry, such as religious materialism (to use the expression of the Tibetan lama Chögyam Trungpa but it can be applied to all religions), scientism, consumerism, and irresponsible ethical and ecological behavior.

Similarly, the story of the ecological movements seems to have been greatly inspired by a certain perception of the Christian narrative. There are other, very different narratives (Buddhism, Hinduism, etc.) that could broaden the scope of this seminar. Is there not a contradiction between opening up to other traditions (such as those of the Amazon) while clinging—at times unconsciously—to a strictly Christian narrative? Our desire for universal understanding is thus reduced and rarefied in a particular form. Islam was the only non-Christian religion represented in our discussion. Why did no others feature? So during the Dialogue there was a strong emphasis on the historic specificity of the convergence of three areas: a certain kind of Christianity, native traditions, and modernity. The question of modernity (especially the acceleration in the transformation of knowledge and techniques) was raised during the first three centuries of Islam—religion and civilization—but obviously in different contexts.

I feel religious phenomena are analyzed in an ideological way when, as Michael and Ted argue, religions disappear but we are witnessing the rise of spirituality in many different manifestations. This difference must be taken into account, if we wish to influence the ecological debate. We must insist on the various dialogic levels between religious forms and the question of Meaning: the latter goes way beyond the former, albeit while underlining its relevance.

The debate between the "modernists and protechnologists" and "anti-technologists and antimodernists" I feel was outmoded. For the "aware" person technology may have a transforming or even transfiguring power. It may lead to living for the Instant (in the Sufi sense of *waqt*) in a dimension that is spiritual, political, ecological, and so on.

Several times I mentioned the eschatological tone of the Islamic message, and I also noted with interest a "pessimistic-speculative" conclusion in this direction from a nonreligious participant, Eduardo, when he alluded to an "imminent *Parousia*." He believes that only a posthuman messianism may be able to change the conditions of our life significantly. I should also like to mention here some elements from my own text: in several passages in the Koran there is a reference to the idea that God could easily make humanity as we know it disappear and replace it with different forms of life—human or otherwise. I quoted to this effect from the Koran (14:19). We must remember that the Islamic tradition teaches us that there is extraterrestrial and extra-human life.

I am increasingly convinced, at the end of this dialogue but also after wider reflection about what is going on in the world, that the crisis is global. It has various symptoms: cosmic, climatic, ecological, and—on a human scale—psychological, moral, and religious. Fundamentalism and terrorism, of a religious, but not only, nature, are particularly sensitive areas. The integral or integrative vision of reality is typical of Islam and is founded on the principle of Oneness—not only divine *stricto sensu*. This principle enables us to give a meaning to the infinite complexity and diversity of life. "Postmodernity," if we wish to give some credit to this term that characterizes the globalized information age we live in, means that we know everything, that pretence and facades are shown up for what they are and that the world's hypocrisy—be it religious, political, academic, or otherwise—becomes increasingly clear. We live in an age of "revelations," and the ecological crisis is indicative of other indicators. Only it can have major irreversible consequences on the future of humanity.

I go along with Bruno Latour in certain peregrinations of his thought, especially when inspired by Whitehead, and I am happy to borrow the expression "panentheism" used by some of Whitehead's disciples: it is completely in line with what the Sufi Ibn 'Arabî (d. 1240) and his school called the "Unity of Being" (*wahdat al-wujûd*). These Sufis do not profess a primary monism, and even less pantheism, but the awareness that everything that is created possesses only a relative ontological degree of existence, which is "borrowed"—to use Ibn 'Arabî's expression—from the only "Real Being," God. The term "cosmotheism," used by Bruno, also fits in with the Sufi/Islamic vision. I can assure that, according to a little-known Muslim teaching, Islam asserts that the animals—an important category of nonhumans—will be judged and raised up, obviously in line with their own level of awareness.

On the other hand, I do not follow his statements when they are too generic—or marked by a specific Christian background, such as: "Where nature enters, religion has to leave," or "When religion encounters nature, one of them has to go." Here I refer the reader to what I wrote in my paper on the Islamic notion of *Fitra*.

How the Dialogue with other religious traditions affected my thinking, what I learned from it, what new methodological perspective aimed at coping with the ecological crisis can be inspired—if any—by this experience?

In considering our topic, I felt a great affinity between the visions of Eastern Christianity and those of Islam. Here I am thinking of the talks given by two speakers in particular: Elizabeth Theokritoff (especially her discussion of the theme of universal praise; her attempt to distinguish the levels of being, and therefore the levels of analysis; and a shared metaphysical view of the purposes of the creation) and Izabela Jurasz (the Fall as only an incident; God as creative "poet" of the universe). We can clearly refer to the historic relations between this form of Christianity and the Muslim world, but I think that here we are actually exploring the issue of constant spiritual features in human experience more deeply. Incidentally, this corroborates my previous personal experience of discussions on comparative mysticism.

22

PART OF CREATION CALLED TO BE GODS: FURTHER THOUGHTS ON THE PLACE OF HUMANS

Elizabeth Theokritoff

The Dialogue was fascinating and extraordinarily intellectually stimulating, leaving me with many ideas and differences of perspective that I expect to be trying to process for years to come. I actually came away with a much greater sense of the need to think seriously about the "nature" of the creatures around us, and most especially of the human being itself. Exploring Orthodox Christian understandings of "nature" in dialogue with "postnatural" approaches would be a demanding project; but perhaps a rewarding one, especially since "postnatural" ideas seem to be reacting to a very different concept of "nature." On the practical level, however, I was left on balance more pessimistic about the chances for agreement on any sort of environmental action on a scale commensurate with the crisis. The reason for these two reactions is basically the same: Even in this group, which was actually relatively homogeneous in terms of cultural background, our discussion revealed enormous gulfs between different ways of perceiving the environmental crisis and its meaning. In other words, it took us straight into the realm of religious and quasireligious conflicts. Pasquale Gagliardi observed at the beginning that "value conflicts" are the most intractable; but I am inclined to think that here we had not even conflicting values, but conflicting pictures of reality. So I see enormous possibilities for dialogue and exploration, enormous difficulties in reaching agreement on desired outcomes and on what we want to "save," and a pressing need for practical environmental action even if those involved have very different motivations; and the tension between these three runs through all my concluding thoughts.

Although the Dialogue was advertised as revisiting questions of the relationship between religion and science, I could never quite understand how this was a major question, given that various religious traditions have been engaged with environmental concerns and environmental science for quite some time. The tension that emerged as problematic was not between religious and scientific approaches, but between theocentric and secular-humanistic approaches to the world in which we live. Does the ecological crisis remind us that we and all creation are ultimately dependent on God? Or does it reinforce a divide between man who "re-creates"—whatever that means—and a world that is to be molded according to his specifications? As someone more used to discussing ecological questions in a theological (usually Christian) context, I was struck by the anthropocentric, or more accurately anthropomonist, tone that prevailed in much of our discussion. For some at least, the outlawed notion of "nature" appeared to have taken with it into exile any idea of a nonhuman realm with its own value and integrity apart from anything that man may have done or might be able to do with it. "God's creation" does of course firmly enshrine that idea; but I frequently felt that in our discussions, the notion of "creation," too, was more anthropocentric than theocentric in content.

In our discussions on the first day, Ted Nordhaus went to the heart of the problem by raising the question of the nature of the ecological crisis and the conflicts involved in it, as a precondition for deciding what should be mobilized to what end. But this question was never properly answered; perhaps it was tacitly accepted as one on which we could not reach consensus. "Mobilization to what?" nevertheless remains the crucial question. Faith undoubtedly motivates people, but not necessarily to join the causes that top the secular agenda. And "If the trumpet sounds an uncertain note, how shall they make ready for battle?" The "uncertainty" of the note, which had previously been inexplicable to me, became deafeningly obvious in the course of the Dialogue. I would suggest that this "uncertain note" has very little to do with supposed uncertainties in the findings of climate science, and almost everything to do with profound confusion over what the "battle" against global warming is meant to be *for*. To be more concrete about the implications of this confusion: there is certainly a Christian motivation to protect the poorest and most vulnerable from environmental catastrophe. But there is no Christian motivation to do this in ways that insulate humans, whether prosperous or would be prosperous, from confronting some hard questions about the priorities and values assumed by our technological civilization. "Virtue exists for the sake of truth," says St. Isaac the Syrian; "truth does not exist for the sake of virtue"—and that includes "ecological virtues." Nor is "saving creation" very helpful as a stated objective, without careful exploration of what it means. If it refers to the divine economy as a whole, then Christians are obviously committed to furthering it—but it makes no particular reference to ecological problems. If on the other hand it refers to an intramundane human project concerned with prolonging the survival of finite creatures, then its relevance to the economy of salvation is not self-evident, and remains to be explored.

Having said all this, I would still want to explore how Christian faith can motivate real efforts to do something rather than nothing. If we start from an agreement in general terms that climate change, in combination with other environmental pressures, poses serious threats to the well-being or even life of humans and many other living creatures on earth, I would suggest two characteristics of Christianity that motivate a response: hope and love. Neither of these provides sufficient criteria for what sort of actions should be undertaken, but they do help clear away some of the excuses for rejecting this or that course of action.

Hope is not something we talked about much, but I believe that its absence is a major factor in the world's present paralysis. If humans are presented with good evidence for a threat that seems unthinkably vast, far beyond our powers to avert, and at the same time rather nebulous, it is a natural if not especially helpful reaction to bury our head in the sand and hope it never happens. Faith in the Creator who wills the salvation of his creation supplies no easy answers, but it removes the disincentive to action, whether at the personal and seemingly trivial level or at the global. We are required to act, to be his coworkers. This synergistic approach, characteristic particularly of Eastern Christianity, facilitates a robust emphasis on human responsibility without letting the weight of that responsibility become crushing.

Love is an obvious motivation to act in the face of any danger threatening others, most especially when the poorest and most vulnerable people are likely to be the main victims. The love to which Christians are called, and that we see as the highest aspiration, is sacrificial—it is not empty or sentimental, but costly. Certainly, there may be occasions when one can benefit others at no cost to oneself; but if every problem we encounter seems to invite a "win-win" solution, we should be alert to the possibility that we are deceiving ourselves. Faced with a problem such a global warming, it becomes clear that, while everyone stands to gain from solving the problem, every solution also requires sacrifice for someone, whether it involves material goods, or income, or choices, or beauty, or a way of life that we are attached to. This needs to be underlined. The vision (nightmare?) of a high-tech, "postnatural" future is not an alternative to sacrifice, but a declaration of who should and should not be asked to make a sacrifice. It decrees that those who have attained or aspire to the most technologically sophisticated and energy-intensive way of life cannot be asked to sacrifice it; but sacrifice can be imposed on nonhuman creatures, and demanded of those humans for whom engagement with nonhuman nature is central to a life worth living. If sacrifices are inevitable, then: what difference might it make if a significant number of people were determined to take on themselves the requisite sacrifice, rather than trying to impose it on someone else? Granted, there would still be substantial disagreements about policy. Especially when we are talking about global problems rather than personal relationships, deciding how to act for the good of others can be a bewildering and frustrating business. But injecting the principle of sacrificial love could help us to recognize that sacrifice, far from being anathema to all but a few saints, is something to which many people feel instinctively drawn as a way of participating in a common cause. One can choose to mock the fact that this often takes the form of token gestures that do little practical good—or one can build on this foundation, challenging people to reveal a heroism that they never knew they possessed. And a renewed sense of solidarity with all human nature, or even all created nature.

We were also asked to consider the potential of religious rituals, which no one actually discussed. Personally, I was hesitant to talk about "rituals" because of the ease with which they are taken out of context and treated as ways of shaping behavior without reference to the faith that is their raison d'être. But one could refer to several aspects of Orthodox Christian liturgical life, notably the Eucharist, the blessing of waters (Figure 22.1) and the blessing services for various objects. All of these speak of a sacramental approach to the world, in which all things, "natural" and man-made alike, are apprehended in their full reality by being seen in relation to God. First of all, however, there is the very fact of worship. Worship indicates that our primary orientation is toward God. It has been suggested that the "dominion" over the earth given to man in the opening chapters of Genesis (before the Fall) has an essentially liturgical quality: man is to lead the offering of praise from all creation. Worship provides a powerful image of the whole body of the community all facing the same direction—toward the Creator. It should be emphasized that this does not mean that members of the community ignore each other and each other's needs, but rather that all relationships are triangular: we relate to other creatures in and through God.

The Eucharist, as has already been noted in Bruno Latour's paper and my own, is an offering of a human product; the staple of human physical life is offered to God in thanksgiving, and received back as a gift of divine and eternal life. Yes, it tells us that such products can be transformed—by God, not by us—so that they manifest the life of the resurrection already in this age; and it also reminds us that such products are still totally the Creator's gift, not something of our own.

Figure 22.1 Encinitas, CA, Fr. Andrew Cuneo, rector of St. Katherine of Alexandria Mission, throws the cross in the water, courtesy of Fr. Andrew Cuneo.

The blessing of waters, by contrast, takes us into an elemental world unmistakably contrasted with human powers to control and shape the world. The prayers of blessing tell us that all things are God's servants (that is not ours), and there follows a list of his benefactions through water: liberating, cleansing, preserving life. The re-creation of man, his rebirth into the death and life of Christ, is the one sacrament accomplished through matter that has not been shaped by humans in any way (originally, baptismal water was meant to be "living," that is flowing). The blessing of waters reminds us that all our cultural edifices rest on an immense infrastructure of "natural" divine benefactions. We contribute nothing to creating these; our task is to refrain from either destroying them, or cornering them so that they are unavailable to the rest of their intended beneficiaries.

Water, the fundamental symbol of God's operation in and through the physical world, is again the medium for the blessing of all sorts of artifacts (houses, cars, wells, etc.). These services of blessing serve to situate our own "creations" within God's creation; they offer the fruits of our technology for the furtherance of his purposes. The blessing prayers are effectively asking that these means of survival should not merely benefit us for a decade or two, but be an opening onto eternal life. In fact, they point us to ways of using and shaping the world that do not involve trying to "gain the whole world" as our own possession, and consequently help us to gain rather than "forfeit our own soul."

In trying to process the many ideas and concepts that were discussed or alluded to, I was struck by several Christian themes with a twist (especially in the work of Nordhaus and Shellenberger, but also in our broader discussions, especially concerning "creation"). The first of the "Christian themes"—without a twist, even—is the insistence that one cannot have a compartmentalized "environmentalism": environmental problems are not separable from concern with the totality of people's needs. This follows directly from thinking in terms not of "man + nature" but of God's creation, but can evidently be reached by other routes as well.

Secondly, we have the theme that the message of the environmental crisis is not negative but positive. The environmental crisis should indeed be seen as challenging mankind not to lower its aspirations, but to rise to its true greatness. Enormous disagreements open up, however, once we start exploring what this greatness consists in. For Christian tradition, it does not lie in a display of power, in our ability to shape and improve our environment even to the point of "creating" a world to our own specifications: it lies in the fact that we bear the image of God and are able to grow into his likeness. If our impact on the world reminds us that we are "as gods," this can be salutary. It invites us to take note of how a God behaves on earth: "He humbled himself, taking the form of a servant." God's love is kenotic: he "empties himself," laying aside all coercive power. Even the creation of the universe can be seen as the primordial self-limitation of God, as he "makes space" for that which is other than himself; but the kenotic character of the Incarnation is unmistakable. God incarnate accepts the limitations of created nature. He suffers tiredness, hunger and thirst; he refuses the temptation to "re-create" stones into bread. To be "as God" is to be on earth as one who serves, who lays down his life for all. An awareness of our exalted calling is thus not opposed to recognizing physical limits; they become an aid to growing to our true stature. Nor is it opposed to seeing the crisis as a call to repentance—*metanoia*, a "change of heart." Indeed, it may be only when we understand our true greatness that we realize how the "goods" that we have been pursuing fall short of it. Our discussions highlighted the important point that the prevailing accounts of human moral responsibility for the ecological crisis are not entirely convincing; something about them fails to ring true. Shifting the emphasis from the moral to the ontological could well be a step forward—though it might well involve revisiting "human nature!" In this connection, it might be worth recalling that the Gospel saying "What does it profit a man if he gain the whole world, and lose his own soul?" has another variant: "What does it profit a man if he gain the whole world, and lose *himself*?" (Luke 9:25).

This brings us to the third Christian theme: liberation, longing for the infinite. Looking at some of the hopes invested in modernization and its attendant technologies as a liberation from the constraints of the world around us, it is hard to miss a secularization of the Christian promise of resurrection to eternal life, of a new earth—in short, of salvation. This sort of secularized eschatology poses an important challenge to Christian thinking. One sort of response can be seen in the contemporary (especially Western) Christian's fear of appearing "otherworldly" and the desire to bring Christian faith "down to earth." But for the Orthodox Christian at least, "coming down to earth"—the Incarnation—is accomplished by God for the sake of "taking up" what is earthly, filling it with himself. Salvation is inseparable from the affirmation that we "look for the resurrection of the dead and the life of the age to come." This means that man cannot satisfy his God-given longing for the infinite in a world limited by mortality; and it might be argued that the ecological crisis is what happens when he tries, when we ask of this world (of which human technological skills are part) more than it is capable of delivering. The perspective of resurrection certainly need not make us uninterested in shaping the world we live in. Technological modifications of our environment to make life livable, and especially to serve the needs of others, are in principle entirely compatible with a focus on the resurrection. Given the recognition that the world this side of death is the field for our work and practical expression of love, such projects can be seen as "little liberations" that point forward to the great liberation from death; and the fact that we keep coming up against physical limits helps to remind us of the difference between the two. But a grand transformative technological project that associates technology with "saving creation" seems to demand a confession that this world bounded by death is the most we can ever hope for. Here technology becomes not an image of salvation but a counterfeit of salvation; and I believe that the human being, and all creation, deserves better than that.

I am very wary of apparent attempts to conflate human technological transformation of our environment with divine-human transfiguration of all creation. The Christian tradition (Eastern, at least) certainly affirms that artificial transformations can be a way of "releasing prayer from things" (Olivier Clément), even as they also enable us to survive; it would be worth exploring Orthodox writers of the nineteenth and twentieth centuries who emphasize the potential of human "creativity," to see where their thought might lead. Human edifices can also express a beauty that images the transcendent, taking us beyond ourselves. But to claim for Christianity a "positive view of all artificial transformations" is a considerable overstatement. Even taking account of the view of some theologians that the nonhuman creation praises God only through us, such a claim that all praise of God on the part of nature involves human intervention is hardly the same as asserting that all human intervention in nature expresses praise of God. Our modern technological culture has gone a long way beyond merely serving survival; and, far from revealing matter as praising its Creator, many of our inventions are more likely to distract us from him and obscure our dependence on him. In such cases, whyever not abandon our "creations" and have the humility to learn from our mistakes? There is every likelihood that new technologies will often be required to redress the harm done by our earlier efforts, but this is not necessarily much to be proud of. Since we heard in the course of the Dialogue from Frankenstein and *The Tempest*, perhaps we might also recall Macbeth on this point: "I am in blood / stepped in so far that, should I wade no more, / returning were as tedious as go o'er."

Turning now to the question of "nature" versus "creation": I am even now somewhat bemused by the arguments over "nature." They seemed to be reacting primarily to a concept of "nature" (as an agency distinct from the workings of God in his creation) very different from anything that the Eastern Christian tradition has ever held. On the other hand, there seems to be a danger of throwing the baby out with the bath water. I do not believe that we can dispense with the belief that created things have a given, though dynamic, "nature"—that they are not self-creating, but reflect a purpose given by their Creator. If we are to shift our focus from "nature" to "creation," it should be made very clear that the "creation" is God's, not ours. It is he, not we, who does the creating—and the saving. But this means that a number of the things apparently deemed objectionable about "nature" have to be affirmed no less clearly in respect of "creation." By all means, let us stop personifying "Nature"—but now we have to deal with the fact that the order in the universe is a personal energy, albeit one ontologically distinct from creation. If we want to "deconstruct" our understandings of "nature" and "the natural," it will have to be in aid of a clearer insight into what the Creator is actually doing in his world—not in aid of telling ourselves new stories better suited to our own current preoccupations. The notion of "nature," whatever its shortcomings, has the virtue of placing man in the context of something vastly larger and more powerful than himself, of inspiring awe and humility. Replacing "nature" with "God's creation" is a step forward, in that the sense of awe is addressed to its proper object. But replacing "nature" with our creation, with a man-made world in which man is the only referent, would seem to be several steps backward.

What methodological approaches might be inspired by our dialogue? I would refer to three points here, two of which I have touched on above in discussing "Christian themes." (1) "Environmentalism" cannot be separated out from the totality of human needs. Helpful concepts here are "God's creation" (which of course includes humans) as well as "the earth as shared habitat." This approach does not ignore conflicts of interest, but it does hold out hope for mutually acceptable compromises. (2) The ecological crisis should occasion an inspirational call to human beings to raise their sights, not lower them. There may indeed be several mutually exclusive understandings of what this amounts to in practice; but the debate over our true calling as human beings is well worth having. (3) On the vexed question of technology and the criteria for evaluating it, a useful term to inject into the discussion might be that of wisdom (I am not advocating the invocation of sophiological philosophies, although some will no doubt want to go in that direction). To see our ability to shape the world as a form of practical wisdom (as did the Church Fathers) is to expect that it should echo, be congruent with, the wisdom in which all things are made, which is Christ. Discovering what this means in a given case will always be an adventure; it cannot be predetermined. But it is certainly not arbitrary, or dependent on individual claims to act on divine authority. The "checklist" for action includes all the ways in which Christ is "embodied": the ethos of the Gospel (love for God and neighbor, the sacrificial love of Christ, the commandments and beatitudes, etc.) and also the "book of nature." "Reading the book of nature" in this context involves trying to discern how wisdom might be operative in the existing order, whether that is a more or less pristine setting or a "natural" response to earlier human intervention (i.e., we are interested in the wisdom operative in the material world, rather than "naturalness" per se). When we look at the "evils" in the created order, what might be their purpose in God's economy? Why are they allowed? How do they compare with the evils and goods that might come of our intervention? "Reading the book of nature" in this way is likely to yield something close to the precautionary principle. Many people will, of course, find the above questions naïve or meaningless. But it is perhaps worth pondering: how would we proceed if we truly believed that everything in creation has a purpose, that is a role in the process of salvation; and that our own purpose involves cooperating in this task? The purpose of things in our environment might encompass human use and transformation; it might involve the needs of nonhuman creatures; it might be (and often is) a function that becomes apparent only after the creature concerned has been removed. Here we have an answer, I suggest, to the question of whether our "natural theology" promotes activism or passivity: it depends on the particulars of each case. Appropriately, perhaps, for the context of a dialogue, I would advocate a "natural theology" that promotes listening.

Defining the Commonplace as the Common Place: Cosmopolitanism and the Cross

Anne Marie Reijnen

Why do I keep such vivid memories of the interdisciplinary meeting held on the island of San Giorgio Maggiore? Why should this particular Dialogue, organized by the Cini Foundation, called *Protecting Nature or Saving Creation? Ecological Conflicts and Religious Passions* stand out so starkly in my mind?

All of us have participated in congresses whose interdisciplinary reach was far more extensive, for example, the annual conventions of the American Academy of Religions; we have known meetings involving participants from more diverse origins—Asia and Africa were entirely lacking in Venice. We may have felt at times that we managed to embody "the conflicts and passions" ourselves; but in my experience, meetings with German-speaking scholars can be even more contentious. Retreating together for a set number of days certainly enhanced the intensity of the exchanges, but again I have participated in similar experiments, notably the interfaith conference about ecology and the spiritual traditions of religions convened by the Ecumenical Patriarch of Constantinople, Bartholomew, in 1998, or the weeklong sessions of the *Groupe des Dombes*, which are truly retreats, as they take place within a monastic (Benedictine) community. I can only explain the enduring spell of the 2010 Dialogue by recognizing that is was unique in having some of all the former, and something else besides. In my recollection, the island becomes a microcosm because it was, for a given time, the focal point of so many views and facts, theories, value judgments, and emotions; and because the beauty of the Laguna, the air and the light, as well as the treasures of music and the peals of the bell tower and of the visual arts were such an integral part of our conversation. Because of the felicitous convergence of natural, aesthetic, spiritual, and cognitive elements, the microcosm of the island became for the time being a *hortus conclusus*.

Looking back now from some distance at San Giorgio, at a remove in terms of both time and of geographical distance, I would like to share a few remarks regarding the difficulties of interdisciplinary conversation, and to elaborate somewhat on the central image in my contribution: the *kosmos*. This will be my personal contribution to the question asked of all the participants of the Dialogue: "Can we mobilize the transformative energies of religions in order to cope with the ecological crisis?" I do believe we can, if we focus on the beauty and the precariousness of our common habitat, an attitude made possible by the vision of the Earth as a discrete entity.

But first a few considerations about the methodology of the dialogue.

THE NATURE OF THE GAME: DECONSTRUCTIVE/ CONSTRUCTIVE

The San Giorgio Dialogue allowed for the encounter of scholars from different disciplines: history of ideas and philosophy, anthropology, geology, and the engineering sciences; theologians representing different Christian confessions and disciplines, for example, Patristics and dogmatics; and a scholar of Islam and of Muslim teachings on "nature" and creation.

In interdisciplinary dialogue, the name of the game implies the exchanging of information and the rule of the game is fairness in listening to the other. More strenuous efforts are needed, probably, to muster equanimity toward others than to assimilate new information. I submit the difficulty has to do with values more than facts. We tend to interpret the opinions of others as the corollaries of the values these others hold. Yet when others proceed with us similarly, we are inclined to reject their reading of us: we feel it is inexact, and probably unfair. "So you are a Christian? Then let me tell you what you believe; be assured, by the way, that all your opinions are flawed, because of your a priori categories." But really how could it be otherwise? It is unreasonable to expect that one partner in the conversation be herself or himself "value-free," and capable of evaluating the other's values in a manner that is both just and compassionate. Interdisciplinary confrontation, then, requires critical and self-critical reflection about values. In other words: deconstruction, followed by (re)construction. In the controversy, one step toward clarification is the unpacking of prejudice regarding, for instance, Christianity. It is achieved by complexification of the mental picture: far from being monolithic, Christianity is a process, an ongoing struggle between dominant and dissident traditions. It is feasible to retrieve relatively marginal strands and practices, such as the Franciscan tradition of kinship with all the elements of "nature," and the spiritual dignity of accepted poverty, or the "Celtic" tradition that seems to have been profoundly attuned to all of the processes of the material world.[1] This retrieval is a means to deconstruct stereotypes, and to (re) construct a more variegated portrait of Christian beliefs.

NEITHER HETERONOMY NOT TOTAL AUTONOMY: THE THEONOMOUS UNDERSTANDING OF HUMAN AGENCY

It is not enough to be critical of one's adversaries' stereotypes: a self-critical stance is equally indispensable. The ecological movement has indicted the harmfulness of the anthropocentric and hierarchical worldview associated with Judeo-Christian values: all of nature is ordained for the benefit of human beings because they are different from and superior to all other living and inanimate creatures. In fact, this is not true of all of the Christian teachings on the topic; already, a careful reading of the main texts from Scriptures—for instance, the creation stories in Genesis 1 and 2—entails a far more complex discourse. Keeping (*shamar* in Hebrew) the earth is not exploiting it. The mandate of the earthling, "to keep" the Garden, is the very same verb as in the expression "to keep the commandments of G.d,"[2] or to put a hedge of thorns around the Thora to protect it. G.d is the "keeper" of Israel (Ps. 121). And Cain asks: "Am I my brother's keeper?" (Gen. 4, 9). The Hebrew verb *shamar* does not primarily mean to "dominate"; it denotes activities such as to preserve, to observe, to be in charge of, and to watch over. Keeping does not imply a conservative approach to "nature conservation." The Swiss have such deep-seated respect for their forests, quasisacralized, that deforestation is outlawed and the Swiss forest is expanding, encroaching on the habitat of many animals who need a mixed environment; the more ecologically responsible attitude is one of human intervention where necessary. But the critique by modern ecology of the Jewish-Christian faiths is not entirely ill-advised or wrong. The Hebrew Scriptures (Tanakh) and the Christian Bible (Old Testament and New Testament) are massively "biased" toward human beings, their doings, and destiny. The anthropocentric view of the world and history, a characteristic of the three "monotheistic faiths" is very widely shared throughout the modern West. For us, to come across an alternative vision, one that is centered on the life of the entire cosmos, is to feel truly disoriented.

A few examples: in the artifacts of so-called primitive cultures, time is shown as cyclical, not linear; as repetitive, instead of nonrepetitive. An Aztec stone calendar (Figure 23.1) represents the world's history, with the sun as its centre, surrounded by symbols of the destruction of past and presents worlds. By contrast, the paradigm prevalent in the so-called common era, shared by the Abrahamic religions and "secular" historiography, supposes the linearity of time and its nonrepetitive nature. You find an expression of the Christian interpretation of history as the unfolding of divine purpose in the following quote: "In many and various ways God spoke of old to our fathers by the prophets; but in these last days he has spoken to us by a Son."[3]

Call it progress ("belief in progress") or revelation ("belief in revelation"): the common trait is the assumption that things have a beginning, and that overall they improve as time goes on. Therefore, the contemporary trust in the unlimited possibilities of technology strikes me as the sibling of Christian hope in providence. If there exists indeed an analogy between progress and ongoing revelation, we should ask ourselves what impact these "beliefs" have in confronting the daunting ecological crisis. Let us say first that is better to hope than to fear. Fear is a bad adviser, since the actions it inspires are reactionary and tend toward repression and exclusion.[4] Hope, on the one

Figure 23.1 Aztec stone calendar found at Tenochtitlan, National Museum of Anthropology, courtesy of Instituto Nacional de Antropologia e Historia, México.

hand, is common to Christians and, on the other hand, philosophers and engineers whose study of the sciences and the history of technology leads them to trust in some man-made *deus ex machina*. For instance, manufactured islands offshore next to the devastated coastal region, a giant shield to protect the overheated planet from the sun's rays; or colonizing some neighboring planets such as Mars, when humanity will finally have made its original biotope unlivable. Christians trust, without necessarily articulating it as clearly, that God will not let go of creation and of God's "favorite" creature, human beings.

"Protecting nature or saving creation," by human agency or divine intervention, refers to an alternative. The dividing line, I believe, runs between autonomous self-reliance—the ideal of secular science and ethics—on the one hand, and theonomous self-reliance on the other hand. Biblical anthropology supposes free will: choose life or choose death. However, the enjoyment of this liberty, and of the fruits of the earth, are not absolute in scriptural terms; they are always ordained toward a goal, the praise of G.d (doxology), the "unconditional" reality under, above, and within all that is conditional. For sure, this framework of gratitude owed to the Giver raises doubts about effective human autonomy within the Jewish and Christian religions. It reeks of heteronomy: of subjection to a metaphysical tyrant—benevolent or not. To fend off this suspicion, believers describe the relationship as "theonomous": not absolutely autonomous, since that would make an idol of the human will, nor heteronomous, since that would encroach on human dignity. "Theonomous" may be a term understood and accepted only within the circle of faith; what it seeks to convey is the existential connection to the Source of good and beauty. The world is not a given ("datum") at our disposal, but a gift for which we give thanks to the Giver, mindful of the commandments that "humanize" our dealings with the other animals, for example, the prohibition on consuming blood or eating the "kid with the mother's milk."[5]

We do not own the earth; we are its stewards. Stewardship involves the responsibility not only for the many who are alive now, but also for the generations to come, not only our children but the children of our children's children, all as yet unborn.[6] Different though the technological and the biblical paradigms may seem regarding self-determination, they are in fact, I believe, closely related. They share an anthropocentric, geocentric view of the universe and they extol activism, be it as "research and development" and/or as theism, the belief in a personal and "active" God. These beliefs are not universal: they are relativized by the existence of radically different worldviews, like the one illustrated by the Aztec stone calendar (Figure 23.1). The Aztec view of the world is heliocentric and in it, worlds are made and are undone as a matter of course.

The scientific approach to reality and the Christian view are connected in yet another important way: inquiry and a critical mindset are essential for both. But, as St. Augustine said, "*Nisi credideritis, non intellegitis*": you do not understand, unless you have believed,[7] quoting the words of the prophet Isaiah. Michael Polanyi has argued that "despite widespread claims to the opposite effect, such is the underlying rationale of modern science."[8] According to Colin Gunton, "Faith however, is not blind faith. For theology, faith does two things at least. It orients us to the place where God is to be found; and it drives us to seek understanding [...] it is noteworthy, although often missed, that it can be seen to apply to other areas of thought and culture than the simply theological."[9]

Thus it is a remarkable, but not surprising fact that in 1663, of the initial members of the Royal Society, a majority of the scientists were Puritans: 42 out of 68, a figure far higher than the small number of Puritans in the overall population should normally yield. As the Harvard sociologist Robert King Merton showed in his study of 1938, the spread of Puritan values encouraged the growth of modern science in seventeenth-century England.

RECLAIMING THE COSMOS AS "THE COMMON PLACE"

The motto "Reclaiming the *kosmos*" entails deconstructive and reconstructive work. The deconstruction is necessary to heal the "theological injury"[10] of the past: parts of the traditional teaching about redemption, especially Protestant teachings, have tended to be "acosmic": indifferent to the common space constituted of land and bodies of water, rocks and vegetation and inhabited by animals—the planet earth—itself within a vast expanse of stellar space with innumerable bodies. This *Kosmosvergessenheit* of large portions of Christian beliefs and practices can be understood as the corollary of the faith in human beings as created in the likeness and resemblance of God (*imago dei*), a doctrine that has wielded immensely beneficial consequences as well. Unfortunately, it has been misused to justify the exclusive relationship between the "soul" or "spirit" of human beings and their God and Redeemer. The other nonhuman animals and the natural world became, relatively speaking, unimportant. In short, the Christian atopia has contributed to the dystopia we know: a world devastated and contaminated through human agency.

I would argue that *kosmos* (cosmos) is a common place in the dialogues about "protecting nature" and "saving creation," since the word is used in different disciplines. In the vocabulary of natural sciences and philosophy, one finds "cosmology," "cosmographer," "cosmotron";[11] in theology we speak of the "cosmic Christ." Here, it seems, we have a "common place" for different types of discourse. The word *kosmos* itself has multiple layers of meaning: "order" or "harmony" (hence: beauty); the universe as an embodiment of order and harmony.[12] It is sometimes used for "world," while the adjective often refers to the universe, exclusive of the earth. Cosmic in a moral sense is synonymous with vast or lofty: "The dark chaotic dullard, who knows the meaning of nothing cosmic or noble." Overstating the consensus somewhat, maybe, William Lane Craig writes: "The traditional doctrine of *creatio ex nihilo* is therefore remarkably consonant with contemporary physical cosmology."[13] I think cosmos is a valuable concept because it refers both to a tangible and to an intangible reality, since "cosmos" is visible and invisible at the same time. I also attach great importance to the political and philosophical ideal derived from it: "cosmopolitanism" (cf. infra).

But the starting point must be existential. What happens to human beings when they can finally see their own world as an object in space? Out of the blue, humanity received a vision of its habitat, seen from outside: the image of the planet Earth, taken from outer space and beamed back to observers.[14]

As I pointed out in my paper, the image has become a picture, since 1972; an "icon," in contemporary parlance. It is now a commonplace, trivialized by its mass reproduction, reduced to one dimension—a flat Earth, as it were—displayed endlessly as "screensavers" of computers and as "wallpaper" on cellphones. We may well ask: has the vision lost its power to astound? When, 40 years ago, human beings saw their planet in this way for the first time, it must have been a memorable moment. G. K. Chesterton pays tribute to the power of first impressions: "In order to strike, in the only sane or possible sense, the note of impartiality, it is necessary to touch the nerve of novelty. I mean that in one sense we see things fairly when we see them first."[15] The subsequent commodification has certainly blurred the initial sense of wonder. Yet the process of trivializing also implies a widespread adoption of the image; by dint of its popularity, it is accessible—and somewhat "cheapened" as a consequence—but it also is universal.

It has occurred to me, since our meeting in September 2010, that representations of the (empty) cross and of the crucifix have likewise proliferated since the birth of Christianity as paintings and sculptures, and as objects in wood or metal. Commodified, the Cross retains the power to comfort and to scandalize. Following up on the similarity between the "object" Earth and the cross, which struck me when reflecting on the Dialogue, I would like to construct (to "invent") an analogy between the image of the Earth "made image by human being," and the image of Jesus of Nazareth on the Cross. The cross and crucifix are artifacts, just like the vision of the Earth from outer space. They were first made at a time and place that can be defined. Both inaugurate an epoch: BC is radically different from AD[16] and the "space age" is a new chapter in the history of *homo sapiens*. At first glance, both images are unsettling: the torture and execution of a human being, and the smallness of the Earth.

But there is a difference. The vision of a crucified man receives additional depth through the words that accompany the image. In fact, for Christian belief, it is only by means of words that the vision is properly understood. A man nailed to a cross or hanging from a tree[17] is a gruesome scene, but alas not an uncommon one. The cross on which Jesus of Nazareth breathes out his last breath acquires unique stature, and a scandalousness of a different kind, only when it is viewed with the words that interpret this event. I submit that in order to "see" the image of a man on a cross as "the Cross," you first need to hear or read the words that come with it. First, the sign on top of the cross, often shown by artists only with the initials INRI, which is short for the Latin title *Iesus Nazarenus Rex Iudaeorum*. In the Gospel according to Luke, we read the following account of the accusation: "And a superscription also was written over him in letters of Greek, and Latin and Hebrew, 'This is the King of the Jews.'"[18] My point is that the empty cross and the empty tomb, the crucifix and the "stations of the cross" have no unequivocal meaning in themselves. They require the interpretation by the early Christian community of the words of the Scriptures (the *Tanakh*, soon becoming "the old Testament" of the Church), for instance, the words of the prophets. For the new "hermeneutical community," the early Church of Christians of Jewish and of Gentile origins, the message of this "image" was one of good news. The ugliest scene one can imagine, for sure, is one that shows the suffering of a human being. It is something that we recoil from: the sight of a fellow human being who is ridiculed, forced to undress, whipped, and tortured. It is only indirectly, by way of paradox, that this scene can become the harbinger of reconciliation. The "hermeneutical community" that gathered around the Scriptures to break bread together "in memory of him" proclaimed to the entire inhabited world, to the *oikoumene*, a message that transformed the single, particular event—the death of a crucified Jew from Nazareth—into something of universal import. The message is, as Church Father Jérôme writes to a young man who wants to become a monk: "*omnes homines posse salvari*": all human beings can be saved.[19]

On first sight, there are many differences with the image of the Earth seen from outer space. First of all, by contrast with the cross that requires words and a community of interpretation to be correctly understood, the vision of the planet requires no words at all to be grasped. It is what is seems; it is just "out there"; you get what you see. But really there is more than meets the eye. This image of the planet taken from a distance represents the apex of the "objective world." According to Philippe Descola, two traits characterize "naturalism," as distinct from the three other ontologies he describes, animism, totemism, and analogism: the belief that human beings are radically different from all other beings because of the specific cognitive capacities and their inner life, while being similar to other beings in terms of their physiology. Hence, starting in the fifteenth century in Western Europe, the keen interest in realistic paintings of individual men and women, and in the realistic rendering of "nature."[20]

As I have shown in my presentation during the Dialogue, the earth had been represented as a globe in Byzantine paintings and sculptures of God as "Pantokrator." In the visions of Julian of Norwich, the earth was like a hazelnut in the palm of the Godhead. In a way then we can say that technology has finally caught up with the fruits of our imagination. In medieval illuminated manuscripts, the earth was shown as a circle surrounded by flames; the heavens above were the abode of the Creator. No one mistook these images for realistic pictures: they were effortlessly decoded as vehicles of a belief. But the same sphere, photographed and beamed back to Earth, as in the "iconic" picture of 1972, has a totally different meaning. For modern people, there is no hierarchy, there are no more angels nor inferior regions. At the same time, the "naturalistic" assumption about the distinctiveness of humankind because of the Cartesian "cogito," is validated. No other animal has contemplated all of its own dwelling from such distance, much less been able to manufacture the device that "records" the image, transmits and receives it, and then to copy and to distribute it. Although human beings are absent from the scene, they are most emphatically active behind the scene, as "demiurgs." In a trivial and humorous way this was expressed by the motto "You are like gods. Better get used to it" accompanying the image of Earth, as Michael Shellenberger and Ted Nordhaus told the participants of the San Giorgio Dialogue.

At the time of the Renaissance and the early industrial revolution, a process was starting whose results we are now reaping. Prior to modern times, for millennia, human beings were exposed to factors that were not of their making: the temperatures on earth, for instance, were a "given." You could emigrate or insulate, but you could not modify the climate. Very recently this has been radically changed. Some experts say that we have entered the era of the "Anthropocene": for the first time in the history of humankind, the weather is subject to human activity. The biosphere is one single huge Pandora's box.

To come back to my analogy: is there a message underlying the picture of the "blue orange" (Figure 7.1) analogous to the message regarding the cross? The cross and the empty tomb require the interpretation by the Scriptures: "For God so loved the world, that he gave his only begotten Son, that whosoever believeth in him should not perish, but have everlasting life."[21] Such is the good news that comes out of the gruesome event of the crucifixion of an innocent. We may ask: is there also good news to be found in what is already good and beautiful (*kosmos*), the picture of the blue orange? I would say that first of all, the "holistic" vision of the planet Earth compels us to recognize the ineptness of dualism and other divisions. Attempts to make sense of the plurality, for instance, by dividing the world politically and economically in "first," "second," "third," and "fourth" worlds, or by opposing developed and emerging nations, or by conjecturing about the destinies of groups of nations (BRICS and PIGS) are ludicrous when put into the cosmic perspective.

The twentieth century was intensely political; indeed, even the conquest of the "last frontier," space, was political; it was a race between Sputniks and Soyuz versus Apollo, and the astronauts were mere pawns in a chapter of the Cold War, as Simon Schaeffer pointed out during the discussion. The twenty-first century will be religious, for sure; and ecological! By which I do not deny the continuing importance of economy and of geo-political factors. As Ted Nordhaus and Michael Shellenberger from the Californian think-tank "Breakthrough" were quite right to point out, the immensely popular image of the Earth seen from sky was the result of complex and costly engineering. It did not happen just like that; it took years of budgeting and investing to reach the sophisticated know-how needed to launch the spacecraft, and to develop the camera that took the picture. On this one planet, huge differences in income and well-being persist. The poor of the underdeveloped and emerging countries pay the higher price for ecological disasters, while contributing far less to the problem, that is emissions; therefore a plea for greater ecological stewardship cannot do without the tools of economy. Yet any ecological catastrophe makes national and class boundaries look like trifles. In spite of the assurances by the French government, the nuclear cloud from the Chernobyl meltdown did not stop at the German-French border. The contaminated waters from the Fukushima reactors that flow into the Pacific spell bad news for the poor and for the rich. The consequences of the heating of the planet ("climate change") will be more painful for the less affluent. In the year 2010, 38 million people were already made "climate refugees" by floods and droughts; most of them were poor.[22] Some consequences will be disastrous for all. Willy-nilly, we are all "cosmopolitans," citizens of the world.[23]

To come back to the analogy between the vision of the cross and the vision of our own Earth: the cross came to manifest a message of universal appeal: "all can be saved." To establish the universalism in a language that could be understood by many, Paul the Apostle looked for a "common ground." He forged a syncretism between the grammar of the Law and the prophets (the Hebrew Scriptures) on the one hand and the "Esperanto" of Greek philosophers, who spoke about the *Logos* on the other hand. In a time saturated with gods, he could point to the Cross and proclaim that it represented the culmination of the unknown gods and of the Lord God known to the people Israel. The common ground was found in history. In the story of Jesus of Nazareth, in his teachings and his passion, "all can be saved."

Today, the common ground is found in the material world. This message is also universal, but it starts out not with the offer of redemption (as in the good news), but rather with the sobering announcement of universal perdition. This is the one place where humankind can live or collectively perish. The vision of the planet Earth seen from outside is akin to an epiphany for me. It is the one "thing" humanity has in common. It is the "common place" to be shared by over 7 billion of women, men, and children.[24] To become aware of the precariousness of this common habitat is an experience shared since the 1970s with countless people of good will, nonreligious and religious alike.

For Christians, there is no escaping the call to "keep" (*shamar*) the earth, the domain of the earthlings. "I call heaven and earth to record this day against you, that I have set before you life and death, blessing and cursing; therefore choose life, that both you and your seed may live."[25] Against extreme anthropocentrism with its hierarchy of being, we now say that human beings and nonhuman animals have the same molecules; physiologically and chemically, there are few differences. Moderate anthropocentrism keeps saying that whether by some superior design, or by a quirk of evolution (emergence), human beings do have some prerogatives: Chesterton identified art as the specific domain of *Homo sapiens*; for me, in the context of this paper, what is most relevant is the capacity to anticipate and to choose.[26] To project and to deliberate are the exclusively human faculties—as far as we know—that explain the cognitive superiority on the one hand, and the moral or ethical potential of the human person on the other hand. It is because of them that we can consider ourselves as *finitum capax infiniti*.

Can we mobilize each other at the brink of a collective disaster? Within limits, we are free!

NOTES

1. An axiom attributed to Saint Columbanus: "If you want to understand the Creator, understand created things."
2. There is a lot to be said in favor of the tradition of considering the name of the Lord as unpronounceable. That is why I will sometimes write "G.d".
3. Hebrews 1:1–2a.
4. Jürgen Moltmann, "Horizons of hope. A critique of 'Spe salvi'," *The Christian Century* (May 2008): 31–33.
5. Acts 15:29 and Deuteronomy 14:21.
6. The ethics of the "principle responsibility," as elaborated by the German philosopher Hans Jonas, imply clear choices regarding the research and development of energy to "fuel" the needs of the expanding world population. Is it right to advocate the increasing reliance on nuclear energy? Nuclear energy is aggressively marketed as "cheap" and "safe"—cheap to produce indeed, but only because the cost in terms of health and environment is borne by the government, that is the taxpayers, rather than by the nuclear industry itself; and "safe" only if one is willing to pretend that the disposal of nuclear waste can decently be transferred to future generations, and if one thinks that major accidents with

reactors are always an acceptable risk. Stewards of the earth will prefer renewable sources of energy, still largely unexplored, such as the power generated by the sun, the wind, the tides of seas and oceans. There might be a promising development from the application of nuclear "cold fusion" (LENR, Low Energy Nuclear Reaction) to the technology of the E-Cat, the electrical catalyzer, as experimented by Andrea Rossi.

7. *De Libero Arbitrio.*
8. Colin E. Gunton, *Christ and Creation* (Grand Rapids, MI: Eerdmans, 1992), 14, referring to Michael Polanyi, *Personal Knowledge*, 266.
9. Ibid.,1 4.
10. The expression was coined by William Storrar, Director of the Center of Theological Inquiry (CTI) in Princeton.
11. *Webster New Twentieth Century Dictionary* (unabridged), 412.
12. See also Rémi Brague, *La Sagesse du monde. Histoire de l'expérience humaine de l'univers* (Paris: Fayard, 2011 [1999]), 35–45.
13. William Lane Craig, "Cosmology," in *Oxford Companion to Modern Theological Thought*, edited by Adrian Hastings (Oxford: Oxford University Press, 2000), 139.
14. There is a parallel Soviet and American history of the exploration of space. We take as our starting point one of the most famous and widely seen pictures, the "iconic" view of the Earth taken by the Apollo 17 spacecraft on December 7, 1972. The shot of the fully lit globe "happened" thanks to a felicitous alignment of Earth, spacecraft, and Sun. There had been other less spectacular pictures previously: Lunar Orbiter was the first spacecraft to capture an image of Earth rising over the lunar limb, in 1966; Apollo 8 and Apollo 11 likewise beamed back pictures of the Earth viewed from the Moon.
15. G. K Chesterton, *The Everlasting Man* (Mineola, NY: Dover Publications, 2007 [1925]), 10.
16. For sure, the computation of AD (anno Domini) takes as its starting point the putative date of the birth of Jesus of Nazareth; but it is only because of the Passion, and (for believers) the resurrection of Jesus confessed to be the Christ, that such importance is ascribed to the birth.
17. See for this theme James Cone, *The Cross and the Lynching Tree* (Maryknoll, NY: Orbis Books, 2011).
18. Luke 23:38. Cf. also John 19:19–20 who mentions title written by Pilate in the three languages, "Jesus of Nazareth the King of the Jews." The text of Mark 15:26 is summary: "And the superscription of his accusation was written over, The King of the Jews." Matthew 27:37 differs very slightly: the soldiers "set up over his head his accusation written This is Jesus the King of the Jews."
19. Saint Jérôme, "Ad Rusticum monachum," CXXV. In *Correspondance*, Tome VII: Lettres CXXI–CXXX (Paris: Les Belles Lettres, 2003), 115.
20. Philippe Descola, "Un monde objectif," in *La Fabrique des images. Visions du monde et formes de la représentation. Paris*, edited by Philippe Descola (Paris: Musée du quai Branly-Somogy, 2010), 73–97.
21. John 3:16 (Authorized King James Version).
22. Cf. Shahidul Haque, regarding the first State of Environmental Migration (2010).
23. For the history of the idea of "cosmopolitanism" and its relevance for today's globalizing world, see the brilliant essay by Kwame Anthony Appiah, *Cosmopolitanism. Ethics in a World of Strangers* (New York: Norton & Company, 2006).

24. Most demographers expect the global human population to stabilize around 10 billions, toward the year 2050.

25. Deuteronomy 30:19 (Authorized King James Version).

26. The "seat" of the capacity to choose and to anticipate can be located in the frontal lobes of the human brain. See the fascinating book by Elkhonon Goldberg, *The Executive Brain. Frontal Lobes and the Civilized Mind* (Oxford: Oxford University Press, 2002).